AI Morality

AI Morality

Edited by

David Edmonds

OXFORD
UNIVERSITY PRESS

OXFORD
UNIVERSITY PRESS

Great Clarendon Street, Oxford, OX2 6DP,
United Kingdom

Oxford University Press is a department of the University of Oxford.
It furthers the University's objective of excellence in research, scholarship,
and education by publishing worldwide. Oxford is a registered trade mark of
Oxford University Press in the UK and in certain other countries

Published in the United States of America by Oxford University Press
198 Madison Avenue, New York, NY 10016, United States of America

British Library Cataloguing in Publication Data
Data available

Library of Congress Control Number: 2023949020

ISBN 978-0-19-887643-4

DOI: 10.1093/oso/9780198876434.001.0001

Printed and bound in the UK by
Clays Ltd, Elcograf S.p.A.

MIX
Paper | Supporting
responsible forestry
FSC® C018072

Contents

List of Contributors

Emma Bluemke's research focuses on a framework for data and information sharing called 'structured transparency'. She has a background in medical physics, privacy-preserving technologies, and machine learning applications for medical imaging. She holds a PhD in Biomedical Engineering from the University of Oxford and is the Research Manager at the Centre for the Governance of AI.

Ruth Chang is the Professor of Jurisprudence at the University of Oxford and Professorial Fellow of University College, Oxford. Her research interests concern the nature of normativity, value, reasons, rationality, agency, choice, conflict and their connections. Her work on choice and conflict has been featured in television, radio, magazines and newspapers globally. She has been a lecturer or consultant for both business and governmental institutions. She is co-editor of two forthcoming volumes, *Conversations in Philosophy, Law & Politics* (OUP), and *Legal Argumentation* (Elgar).

Gabriel De Marco is a Research Fellow at the Oxford Uehiro Centre for Practical Ethics at the University of Oxford. His current research focuses on free action, moral responsibility, and the ethics of influence.

Thomas Douglas is Professor of Applied Philosophy and Director of Research at the Oxford Uehiro Centre for Practical Ethics, Faculty of Philosophy, University of Oxford. Thomas trained in clinical medicine and philosophy and works mainly on ethical issues raised by medicine and the life sciences. He is currently leading the project 'Protecting Minds: The Right to Mental Integrity and the Ethics of Arational Influence', funded by the European Research Council.

David Edmonds is a Distinguished Research Fellow at the Oxford Uehiro Centre for Practical Ethics. He is the author or co-author of many critically acclaimed books including *Wittgenstein's Poker* (with John Eidinow), *Would You Kill The Fat Man?*, *The Murder of Professor Schlick*, and *Parfit: A Philosopher and His Mission To Save Morality*. With Nigel Warburton he co-hosts the popular philosophy podcast, Philosophy Bites.

Linda Eggert is an Early Career Research Fellow in the Philosophy at Balliol College and the Institute for Ethics in AI at the University of Oxford, where she works on moral and political philosophy, and the philosophy of law. Her work focuses on issues in non-consequentialist ethics, theories of justice, the ethics of war and defensive harming, and the relationship between AI, human rights, and democracy.

Binesh Hass is a postdoctoral research fellow at the Oxford Uehiro Centre for Practical Ethics and a junior research fellow at Wolfson College, Oxford. His work encompasses topics at the juncture of jurisprudence and applied philosophy, including questions in the theory of reasons (and reasonableness), capacity, and discrimination.

Saffron Huang is co-founder and co-director of the Collective Intelligence Project, and a fellow at the Centre for the Governance of AI in Oxford. She has been a research engineer at DeepMind, and worked on tech governance research and policy with organizations including the Ethereum Foundation, the Harvard Berkman Klein Center, and the British Foreign Office.

Theodore M. Lechterman is Assistant Professor of Philosophy at IE University, where he holds the UNESCO Chair in AI Ethics and Governance. His work spans political philosophy, applied ethics, and the intersections between them. He is the author of *The Tyranny of Generosity: Why Philanthropy Corrupts Our Politics and How We Can Fix It* (Oxford University Press, 2022). Lechterman was trained at Harvard and Princeton and completed postdoctoral fellowships at Stanford, Goethe, Hertie, and Oxford.

Muriel Leuenberger is a Postdoctoral Research Fellow at The Center for Ethics and the Philosophy Department of the University of Zurich. Her research focuses on authenticity, identity and narratives in the ethics of technology and medical ethics (neuroethics in particular).

Maximilian Kiener is at the Institute for Ethics in Technology at Hamburg University of Technology. He specializes in moral and legal philosophy, with a particular focus on consent, responsibility, and artificial intelligence. He is also an Associate Member of the Faculty of Philosophy at the University of Oxford.

Silvia Milano is a Senior Lecturer in Philosophy at the University of Exeter and the Exeter Centre for the Study of Life Sciences (Egenis), and an affiliate member of the Governance of Emerging Technologies programme at the Oxford Internet Institute. Her work explores the epistemological and ethical aspects of data and AI, with special focus on recommender systems.

Peter Millican is Gilbert Ryle Fellow and Professor of Philosophy at Hertford College, University of Oxford, and Visiting Professor at the National University of Singapore. Until 2005 he was Senior Lecturer in Computing and Philosophy at Leeds University, and has spent much of his career working in that interdisciplinary space, focusing on the development and teaching of joint degrees, software for outreach, and research on Alan Turing and the philosophy of AI. He has also published more than forty papers on early modern philosophy, especially the work of David Hume and related topics in epistemology, ethics, and philosophy of religion.

César Palacios-González is the Deputy Director of the Master of Studies in Practical Ethics, at the University of Oxford; a Senior Research Fellow in Practical Ethics at the Oxford Uehiro Centre for Practical Ethics; and a Research Fellow at Wolfson College. His research interests include philosophy of medicine, practical ethics, and AI ethics.

Carina Prunkl is a Postdoctoral Research Fellow at the Institute for Ethics in AI and a Junior Research Fellow at Jesus College, Oxford. Her research focuses on human autonomy and fairness in the context of algorithmic decision-making. She also works on responsible research and the governance of AI. Carina Prunkl previously completed a D.Phil. in Philosophy and an MSt in Philosophy of Physics at the University of Oxford, as well as a BSc and MSc in Physics at Freie Universität Berlin.

Jonathan Pugh is a Senior Research Fellow and Manager of the Visitors Programme for the Oxford Uehiro Centre at the University of Oxford. He is also an Official Fellow in Ethics and Values at Reuben College. His research interests lie primarily in issues concerning personal autonomy in practical ethics, particularly topics pertaining to informed consent. He has also written on the ethics of insurance, criminal justice, neuroethics, and gene-editing.

Daniel Susskind is a Research Professor in Economics at King's College London, and a Senior Research Associate at the Institute for Ethics in AI at Oxford University. He is the co-author of the best-selling book *The Future of the Professions* (2015) and the author of *A World Without Work* (2020), described by the *New York Times* as 'required reading for any potential presidential candidate thinking about the economy of the future'. Previously he worked in the Prime Minister's Strategy Unit, in the Policy Unit in 10 Downing Street, and in the Cabinet Office.

Divya Siddarth is the co-founder of the Collective Intelligence Project, an experimental research organization that advances collective intelligence capabilities for the democratic and effective governance of transformative technologies. She was formerly Associate Political Economist and Social Technologist at Microsoft, and also holds positions as the Ethics in AI Institute in Oxford and at the Plurality Lab at Harvard's Safra Center for Ethics.

Aksel Sterri is a Postdoctoral Researcher in Philosophy at the Faculty of Health Sciences at Oslo Metropolitan University and The Oxford Uehiro Centre for Practical Ethics. He works in moral philosophy and particularly interested in how markets can be designed for the common good.

John Tasioulas is Professor of Ethics and Legal Philosophy at the University of Oxford and the Director of the Institute for Ethics in AI. He works on moral, political, and legal philosophy. He is the co-editor of *The Philosophy of International Law* (Oxford University Press, 2010) and the editor of *The Cambridge Companion to the Philosophy of Law* (Cambridge University Press, 2020).

Andrew Trask is a PhD student at the University of Oxford and a Senior Research Scientist at DeepMind studying privacy, AI, and structured transparency. He is also a member of the United Nations Privacy Task Force, raising awareness and lowering the barrier-to-entry for the use of privacy preserving analytics within the public sector.

Charlotte Unruh is a Lecturer in Philosophy at the University of Southampton. She works in moral philosophy and is interested in the philosophy of harm, future generations, and the ethics of artificial intelligence, especially the future of work.

Carissa Véliz is an Associate Professor in Philosophy at the Institute for Ethics in AI, and a Fellow at Hertford College at the University of Oxford. She is the recipient of the 2021 Herbert A. Simon Award for Outstanding Research in Computing and Philosophy. She is the author of *Privacy Is Power* (an *Economist* book of the year, 2020) and the editor of the *Oxford Handbook of Digital Ethics*. She advises private and public institutions around the world on privacy and the ethics of AI.

John Zerilli is a philosopher with particular interests in cognitive science, artificial intelligence, and the law. He is a Chancellor's Fellow (Assistant Professor) in AI, Data, and the Rule of Law at the University of Edinburgh, a Research Associate in the Oxford Institute for Ethics in AI at the University of Oxford, and an Associate Fellow in the Centre for the Future of Intelligence at the University of Cambridge. His two most recent books are *The Adaptable Mind* (Oxford University Press, 2020) and *A Citizen's Guide to Artificial Intelligence* (MIT Press, 2021).

Introduction

David Edmonds

Beware. It's upon us. It's not the future, it's the present.

Just a few years ago, discussion about AI was confined to computer scientists and policy wonks. Now there is barely a day without an AI headline in the newspapers or broadcast media. As I write, a group of leading technologists have published a letter warning that AI could wipe out the human race and proposing a range of regulatory measures in an attempt to mitigate the risk.

It would be a mistake to dismiss this as hyperbolic doom-mongering—one should surely take seriously fears which are expressed by not just one or two experts but many. Back in the 1980s there was already a consensus among climate scientists and meteorologists that the world was heating up and humans were largely responsible. These experts were largely ignored—and now the crisis has deepened. Best not to repeat this error.

However, even if the existential danger to humanity is overblown, what can't be gainsaid is the ever-increasing presence of AI in almost every aspect of life—from the economy to law, health, transport, education, defence, media/communication, sport, and leisure. AI is changing our lives in mundane ways, and in ways that are, or will be, transformative. While you will be aware of some of these effects, others are more opaque.

This is not a book about imagining extreme cataclysmic threats from AI. Rather, it is about the ethical challenges that we already see taking shape in different areas of our lives. It is not intended to be comprehensive—there are scores of other AI topics that we could have covered. But this book will, I hope, provide a useful snapshot of the types of concerns that AI is forcing us to confront.

The term 'artificial intelligence' is used in various ways. In essence, though, it entails the performance of tasks by computing systems that previously would have needed human brain power. Of course, for a long time, modern calculators have been able to add, subtract, divide, and multiply, with far greater speed and accuracy than most humans. One new facet that characterizes AI, however, is the ability to 'learn', so that performance on tasks can improve as more information is inputted over time.

Perhaps this is best illustrated with an example. In 1997, in what was a depressing turning point for some chess players, the then world champion, Garry Kasparov, was beaten by the IBM-created Deep Blue computer. Hitherto, chess had been regarded as too unfathomably complex to be mastered by machine. Deep Blue was programmed with advice from strong chess players, but effectively it relied on brute computational power.

However, with the advent of machine learning, chess programmes became even stronger. A new generation of programmes 'learnt' chess by identifying correlations and patterns in chess, allowing the computer to predict what moves are most likely to be successful. The staggering achievement of AlphaZero, a programme from the AI company DeepMind, was to defeat a mighty chess engine, Stockfish, just a few hours after it had been provided with nothing but the rules of the game. AlphaZero generated a huge data set by playing itself millions of times.

Chess is just a game (if one that is taken rather seriously by some of us). But AI is impacting more important domains. In what follows, you'll read about AI and the rise of autonomous weapons, the growing threat of cyber-attacks including in surprising areas such as health, and the use of AI predictions in the legal and criminal system. You'll read too about how AI might change politics, affect our ability to master and maintain various skills, and perhaps cost us our jobs.

If you think you can opt out of these issues, or that they do not concern you, think again. AI functions with data. Chess algorithms are in that sense like all AI algorithms. AI identifies patterns that are often blind to humans and makes predictions and inferences on the basis of past information. The more information algorithms can gobble up the better (for them at least!). You, the reader of this book, are unlikely to live in a cave. Just as slugs and snails leave a trail of slime wherever they go, so anybody who navigates through the modern world leaves a trail of data. We leave data when we visit a doctor or when we 'like' a post on social media. We leave data when we pay for a train journey with our travel card. We leave data when we purchase a product in a shop or online. This, in turn, raises another set of moral dilemmas. Who owns this data? What are the implications for privacy?

*

A number of themes crop up in the book. One is autonomy. AI promises to help satisfy our preferences. For example, an online bookseller, or movie streaming service, will use AI to predict what books or films you might like. Indeed, it might be fantastic at this—you might love all its recommendations. Still, questions remain about whether this is an entirely positive development and, in particular, whether AI may erode the self-control we have over our lives.

A second theme is bias. This is becoming a more familiar worry. AI seems to hold out the promise of overcoming bias: after all, computer judgements, unlike human judgements, are not affected by whether the machine has just consumed a heavy lunch, or quarrelled with its spouse. In fact, rather than AI overcoming human irrationality and prejudice, it may actually entrench it.

A third theme is responsibility. As we subcontract more and more decisions to AI, we face the tricky issue of whom to hold accountable for them. Accountability is normally linked to praise and blame, award and sanction. There's not much point blaming the algorithm, and algorithms can't be punished. So where does responsibility lie? This is a particularly thorny problem in cases where humans do not fully comprehend *why* an algorithm has reached the judgement it has.

A fourth and related theme is privacy and transparency. Governments, companies, and other individuals have the ability to find out much more about us than they ever could in the past. But what information about ourselves should we prevent others from accessing? And if there are suppositions about us based on our personal data, should we not at least be told how these inferences have been made?

A fifth theme is meaning. AI promises (threatens?!) to make many current human activities redundant. This includes work. It's becoming ever clearer that many jobs will soon be done more quickly, more cheaply, and much better by algorithm and machine. Employment, for lots of people, is a vital source of self-respect and fulfilment. If work begins to dry up, it's worth reflecting on whether there might be other outlets to fill a meaning-vacuum.

A sixth theme is the broad one of value and morality. Some philosophers and computer scientists believe that we're approaching the time when we have to recognize AI as having ethical status. This is hugely contentious. But there's a much more immediate and practical matter of how we prevent AI from acting in ways that are alien to human values.

We all care about these aspects of our lives. The rise of AI will force us to reflect on them more deeply and to revise our attitudes and our understanding of their significance. This book, I hope, shows how it can be stimulating and rewarding to think about the moral challenges of the AI revolution.

*

A few thanks.

My thanks to the OUP editorial staff, especially Peter Momtchiloff and Imogene Haslam, to the copyeditor Edwin Pritchard and production coordinator Vasuki Ravichandran.

My thanks to all my brilliant contributors. Each of them has had some connection or other with the ancient university of Oxford, which has been impressively quick to embrace the contemporary topic of artificial intelligence. There are several institutions within Oxford that house researchers dedicated to examining issues thrown up by AI. I have relied for this book on two institutions in particular—the Uehiro Centre for Practical Ethics, my academic base, and the newly formed Institute for Ethics in AI. I would like to thank the directors of both institutes, Roger Crisp (Uehiro) and John Tasioulas (IEAI). John's early enthusiasm for this project was especially important—and he introduced me to several of those who've authored chapters.

PART I
DEFENCE, HEALTH, LAW

1

Autonomous Weapons Systems and Human Rights

Linda Eggert

A few years ago, now retired Special Operations Chief Eddie Gallagher stood accused of horrific war crimes: shooting at unarmed civilians—teenage girls and an elderly man—and pitilessly executing a captured half-conscious teenage fighter. His platoon of Navy SEALs had once revered him. But, after months of vicious fighting in Iraq, they broke the code of silence. As one member of Alpha Platoon, who had served under Gallagher, put it in an interview: 'How am I supposed to teach my kid [the difference] between right and wrong and look him in the eye, if I'm not doing everything that I can?'[1]

*

Autonomous weapon systems (AWS) are standardly defined as systems that, once activated, can identify, select, and engage targets without human involvement. Free from human limitations, they promise the prospect of a world without abuses like those Gallagher committed. They do not succumb to anger or fear or vengefulness. And they can process vast amounts of information at super-human speed. But, also unlike humans, they have no conscience to wrestle with.

The notion of delegating life-and-death decisions to machines has given rise to numerous moral, political, and legal concerns. This chapter examines one of the most fundamental but rarely scrutinised objections: that delegating life-and-death decisions to machines poses an indefensible affront to human rights and dignity.

Whether the use of AWS is consistent with human rights may seem like an odd question. The idea of outsourcing life-and-death decisions to algorithms strikes many people as so obviously wrong that one might doubt whether there is really anything here to be explained. But we need to know whether we can trust our instincts. After all, some objections to AWS might turn out to be contingent—dependent on circumstances. And they might weaken, and perhaps altogether disappear, as these circumstances change.

Take the worry that AWS may not be able to comply with the laws of armed conflict. Perhaps sufficiently advanced AI, with sufficiently advanced contextual judgement, will one day be able to distinguish between legitimate and illegitimate targets, and to make proportionality judgements—correctly assessing, for example, that a munitions factory is vital enough to an unjust enemy to make it a legitimate target, even if the attack would also harm some civilians.

Or consider the worry that eliminating human involvement will create accountability 'gaps' and make it impossible to hold anyone meaningfully accountable for harms caused by AWS. If a human operator activating an AWS was aware that it might cause wrongful harms, we might be able to hold her accountable for recklessly deploying that AWS. If the human operator was *not* aware of this risk, we might be able to hold her accountable for negligently deploying that AWS. Besides, sometimes people incur duties of compensation, even if they are not personally responsible for the harm to be compensated, as with reparations for historical injustice. All this is to say that certain accountability obstacles may not be as insurmountable as they appear.

If key objections to AWS may weaken as circumstances change and technologies advance, we need to know whether objections based on human rights will persist. And to answer that, we need to know what human rights objections amount to.

That removing human decision-makers is somehow 'dehumanising' is a natural enough assumption. The difficulty with concerns about human rights—and dignity, which is often mentioned in the same breath—is that, although appeals to these notions are ubiquitous, it's rarely clear what they mean. They typically amount to little more than placeholders for substantive argument, leaving it obscure *why* AWS threaten to pose a special affront to dignity and human rights.

To some extent, this is understandable. Some protean concepts are like a slippery bar of soap, escaping our grasp as we try to strengthen our grip. But there is too much at stake to be vague. If proponents of AWS are right, these technologies might significantly reduce the number of innocent deaths. Not only are robots immune to exhaustion, panic, rage, and vengefulness, which has led human soldiers like Gallagher to commit unspeakable moral crimes. AWS also save militaries from having to send soldiers into harm's way. And AI's superhuman speed and information-processing power will allow AWS to make split-second decisions that could save lives in situations in which the human mind is simply too slow to process all relevant information. AI may enable a more careful selection of targets and more precisely timed attacks.

Grappling with the question of whether using AWS necessarily violates human rights is thus not merely a philosophical exercise. If we are to reject a technology that could save innocent lives, we had better be sure that our reasons for doing so are sound ones. Feeling a vague sense of unease isn't good enough. If we appeal to human rights and dignity, these notions had better do substantive moral work.

Human Rights

Whether AWS violate human rights, one might claim, is neither here nor there because wars involve countless rights violations. Besides, at least as far as the legal regulation of conduct during armed conflict is concerned, the priority has typically been to minimise harm, rather than to protect individual rights. That's why conduct during armed conflict is governed by international humanitarian law (IHL) rather than international human rights law (IHRL). So, one might say, what matters is not human rights, but whether AWS would have the capacity to comply with the principles enshrined in IHL.

Complying with IHL, however, is only part of the bigger picture. AWS may be deployed in many contexts outside of war—including in hostage situations, going after big-game poachers, and in domestic law enforcement. For example, in a hostage situation in which speed is of the essence and a hostage-taker is only exposed for a split second, an AWS with facial recognition capacities might be able to release force much more quickly and precisely than a human sniper.[2] In addition, AWS could patrol and protect high-security prisons and borders, and they might be used in the 'war on drugs', and in antiterrorism operations. Of course, to point out that AWS *could* conceivably be used in these contexts isn't to say that they *should* be used. The point is that, since the use of AWS may not be limited to armed conflict, their capacity to comply with IHL is not all that matters. The demands imposed by human rights are no less vital.

Considering the use of force against humans, the most directly affected rights are those to bodily integrity, including the right to life, the right to security, and perhaps also the right not to be subject to cruel, inhuman, or degrading treatment. Article 3 of the Universal Declaration of Human Rights (UDHR) states that 'Everyone has the right to life, liberty and security of person.' Article 5 states, 'No one shall be subjected to torture or to cruel, inhuman or degrading treatment or punishment.'[3] The UDHR is not legally binding, but similar principles are enshrined in, for example, the International Covenant on Civil and Political Rights, which states that 'Every human being has the inherent right to life. This right shall be protected by law. No one shall be arbitrarily deprived of his life.' And, according to Article 9(1), 'Everyone has the right to liberty and security of the person.'[4]

Now, one strategy would be to try to show that automated decisions are necessarily 'arbitrary'; and that automated life-and-death decisions necessarily violate the right not to be arbitrarily deprived of one's life. For example, Christof Heyns, the former UN Special Rapporteur on extrajudicial, summary, or arbitrary executions, has claimed that there is 'an unspoken assumption' in IHRL, which demands that decisions to use lethal force must be reasonable—presupposing the necessary capacities for human reasoning—and must be made by a human agent.[5]

But appealing to an 'unspoken assumption' is not entirely satisfying. To find what lies behind this assumption we must look beyond the existing legal human rights framework.

However, a difficulty with asking what human rights *exist*—for example, whether there is a human right against automated decision-making in certain contexts—is that there is a great deal of disagreement about what human rights *are*. This is not the place to try to settle this question; but here's a common starting point. Human rights are universal moral rights that people possess in virtue of their humanity. This appeal to 'humanity' presupposes a set of intrinsic properties that grounds some requirement of respect for people as individuals.

Dignity

'All human beings', according to Article 1 of the UDHR, 'are born free and equal in dignity and rights.'[6] In its Preamble, the UN Charter similarly reaffirms 'faith in fundamental human rights, in the dignity and worth of the human person.'[7]

Arthur Schopenhauer labelled dignity 'the shibboleth of all the perplexed and empty-headed moralists who concealed behind that imposing expression their lack of any real basis of morals.'[8] Most people, the philosophical mainstream included, converge in understanding dignity as some sort of worthiness of respect—an idea that has real consequences.

In 2006, the German Constitutional Court ruled as unconstitutional legislation that would have allowed the Minister of Defence to authorise the shooting down of a civilian airplane involved in a 9/11-style terrorist attack. Although this proposed legislation potentially could have saved lives, the Court rejected it on the grounds that it would have violated the dignity of those on the plane.[9]

The appeal to dignity is revealing. Painting in the broadest of strokes, we might say that dignity is violated when people are treated as less worthy of concern than they really are. This can happen in at least two different ways. The first is relational. We mustn't treat some human beings as worth less than others. The second is based on the related, broadly Kantian, idea that individuals have an 'inner incomparable worth' and what we might describe as immunity from fungibility. We mustn't reduce people to numbers and treat them as interchangeable, disregarding what the political philosopher John Rawls called the 'separateness of persons'.

According to Heyns, 'When someone comes into the sights of a computer, that person is literally reduced to numbers: the zeros and the ones of bits.'[10] But when a human makes a decision to attack a target, things are not necessarily relevantly different: humans can also treat one another as mere numbers. Threats to dignity arise in cases in which people are not given the concern they deserve. Perhaps the real issue is with *how* the decision is made, rather than with whether it is made by a person or a machine. The threat to dignity lies in merely following an algorithm. People can do this just like machines.

Of course, this doesn't alleviate worries about machine-executed algorithmic decision-making. That humans don't always give each other the concern they deserve is no response to the objection that machines cannot give people the concern they deserve.

Now, our focus here is with what dignity requires. Can it ground an objection to AWS even if AWS might ultimately reduce innocent deaths? If each person matters, to borrow Heyns's phrase, 'as a separate and unique, or irreplaceable, individual', then her right not to be killed must be respected, even if violating it could prevent a greater number of such violations. So, even if AWS might reduce violations of certain rights, this is not all that matters.[11] We might still have rights-based reasons to object to their use.

But what if we want to resist the assumption that it is 'in virtue of our humanity'—rather than, say, political and legal agreements—that we have human rights? After all, different people in different cultures might have wildly different views of what, if anything, makes us distinctly 'human'. And, we might think, we can't possibly, and shouldn't have to, first identify some special, universally accepted value-conferring property in humans to be able to explain what, if anything, is distinctly troubling about AWS.

To be sure, respecting people's dignity can take different forms. And this is not the place to resolve protracted debates in the philosophy of human rights. In the end, what matters—rather than some contentious positive claim about what is special about humanity—is a duty not to subject people to dehumanising or 'inferiorising' treatment that denies them the moral concern they are owed.[12]

Taking Stock

We should not be nonchalant about the potential to save lives. One view is that, if AI technology can help reduce civilian casualties by making targeting in the fog of war more accurate, it should be used, but in an 'advisory' capacity.[13] Human decision-makers should use the technology to make better-informed decisions and minimise harm. Indeed, to the extent that it is possible to reap the potential benefits of AI (to increase accuracy and reduce casualties), without eliminating human decision-makers, this straightforwardly seems like the right thing to do.

The problem is that some of the key promises of AWS seem inextricably intertwined with the elimination of human involvement. After all, one asserted advantage lies in the very possibility of replacing humans who would otherwise be made to risk their lives. Another lies in AI's superhuman speed and computing power. Both arguments presuppose that the main benefits arise from humans being removed from the process.

But this is not the only trade-off. One prominent worry about AWS, recall, is that delegating life-and-death decisions away from humans will create gaps in which neither humans nor robots can be meaningfully held accountable for harms caused by AWS. Suppose that eliminating human decision-makers does indeed jeopardise the possibility of accountability; and also continue to assume that AWS may indeed reduce civilian casualties.

How should we weigh the promise of AWS to reduce harm to innocent people against the value of accountability? Should we insist that the importance of accountability outweighs that of reducing harm? Is a world in which more harm is caused, for which people can be held accountable, better than a world in which less harm is caused, but without the possibility of accountability for rights violations?

A plausible response is that, no matter how great the importance of accountability, the ultimate aim must be to prevent harm. This is where proponents of AWS may have a point. The moral imperative to prevent harm to innocent people may well outweigh reasons to retain human involvement for the sake of accountability.

But what our discussion has revealed is that this is not the whole story. There might be distinctly non-instrumental reasons for humans to make each and every life-and-death decision: reasons that have nothing to do with furthering extrinsic values like compliance with the law or the possibility of accountability, and everything to do with people's intrinsic worth as rights-holders, which demands a kind of moral concern that is foreign to algorithms— whether they are executed by humans or machines.

We have now arrived at what might turn out to be the grounds for a right not to be subjected to algorithmic life-and-death decisions. This is striking. After all, insofar as AWS really could reduce harm, their use might be in people's interests.

But this is merely an apparent—and not a real—paradox. Certain rights matter not just because they serve to protect important interests, but also—perhaps even more so—because they protect the intrinsic moral worth of people. So, in justifying a right against algorithmic life-and-death decisions despite the life-saving potential of AWS, what matters is that such a right is valuable not necessarily because of the positive consequences of complying with it, but primarily because of what it says about people as rights-holders.

For proponents of AWS, who treat technology's life-saving potential as paramount, *not* delegating life-and-death decisions to machines on these grounds may seem like too high a price to pay. For those who see objections to AWS besides those that might ultimately turn out to be contingent, it may be a price worth paying.

Conclusion

Is the use of AWS consistent with human rights? If proponents of AWS are right, using this technology may significantly reduce harm. At the same time, the idea of delegating life-and-death decisions to algorithms strikes many of us as troubling.

This chapter elucidated how the idea of human rights might substantiate the intuition that there is something distinctly wrong with delegating decisions to use lethal force to automated processes. The appeal to human rights matters, because other objections to AWS may weaken as circumstances change. We need to know what moral objections, if any, will persist, even if the technology advances.

Non-contingent objections to AWS rest on the assumption that using AWS fails to give people the moral concern they deserve. This isn't because there is anything morally mysterious about machines. Purely algorithmic decision-making violates human dignity, regardless of whether the algorithm is executed by a person or a machine.

Besides, it follows from people's immunity from fungibility that their rights persist even if violating them might reduce the number of rights violations overall. The prospect of reducing harm does not, therefore, automatically justify the use of AWS.

The moral foundations of human rights can play a role that is both philosophically defensible and helpful in answering a practical moral question about AWS. Human rights may ground a distinctly non-contingent objection to AWS. The requirement to treat people as deserving of moral concern goes to the very core of who we are as inhabitants of the moral universe.

The conclusion is not merely that replacing human combatants, like Gallagher, with AWS would do nothing but replace human violations of human dignity with AI-powered violations of human dignity. Rather, it is that our reasons for objecting to both types of violation are the same.

Notes

1. CBS Sunday Morning, 'Eddie Gallagher and the Changing Story of a Death in Iraq' (2022) (<https://www.youtube.com/watch?v=GDARo8VPsLk>). See also David Phillips, *Alpha: Eddie Gallagher and the War for the Souls of the Navy SEALS* (Penguin Random House, 2022).
2. Christof Heyns, 'Human Rights and the Use of Autonomous Weapons Systems (AWS) During Domestic Law Enforcement', *Human Rights Quarterly* 38 (2016): 350–78, 358–9.
3. Universal Declaration of Human Rights, proclaimed by the General Assembly, Resolution 217 A (III), A/RES/3/217 A, 10 December 1948.
4. International Covenant on Civil and Political Rights, adopted 16 December 1966, General Assembly resolution 2200A (XXI) (entered into force 23 March 1976, in accordance with Article 49).
5. Heyns, 'Human Rights and the Use of Autonomous Weapons Systems', 362.
6. Universal Declaration of Human Rights, proclaimed by the General Assembly, Resolution 217 A (III), A/RES/3/217 A, 10 December 1948.
7. United Nations, *Charter of the United Nations*, 24 October 1945, 1 UNTS XVI, Preamble. Article 5 of the African Charter on Human and Peoples' Rights similarly references the notion of dignity, though principally in the context of the right against cruel, inhuman, and degrading treatment; see <https://www.achpr.org/legalinstruments/detail?id=49>. Article 1 of the EU Charter of Fundamental Rights, meanwhile, states: 'Human dignity is inviolable. It must be respected and protected.' See <http://www.europarl.europa.eu/charter/pdf/text_en.pdf>.
8. Cited in Michael Rosen, *Dignity: Its History and Meaning* (Harvard University Press, 2012), 1.
9. BVerfG, Judgment of the First Senate of 15 February 2006, 1 BvR 357/05, paras 1–156, <http://www.bverfg.de/e/rs20060215_1bvr035705en.html>. Also recounted in Heyns's discussion of what he describes as a 'right to dignity' in 'Human Rights and the use of Autonomous Weapons Systems'.
10. Heyns, 'Human Rights and the Use of Autonomous Weapons Systems', 370.
11. Heyns, 'Human Rights and the Use of Autonomous Weapons Systems', 369. Using AWS is not necessarily an instance of violating one right to minimise violations of the same right. More precisely, it might mean violating a right against automated decision-making to minimise violations of the right not to be killed.
12. See Andrea Sangiovanni, *Humanity Without Dignity: Moral Equality, Respect, and Human Rights* (Harvard University Press, 2017).
13. For example, Peter Asaro, 'On Banning Autonomous Weapon Systems: Human Rights, Automation, and the Dehumanization of Lethal Decision-Making', *International Review of the Red Cross* 94 (2012): 687–709.

Further Reading

Dan Saxon, *Fighting Machines: Autonomous Weapons and Human Dignity* (University of Pennsylvania Press, 2021).

2

Cyber-Risks and Medical Ethics

Maximilian Kiener

Paul Pugsley must have trembled when he looked at the computer screen. Paul is an emergency resident at Maricopa Medical Centre in Phoenix, Arizona, and was treating a patient for an acute stroke. To determine the appropriate treatment, Paul had to conduct a CT scan. Suddenly, the computer froze and a message flashed up: 'You have been hacked. Pay or we will shut down your system!' Valuable time elapsed and help came too late. Paul's patient suffered serious brain damage.[1]

Luckily, this was a just a trial run, and Paul's patient was a dummy. Yet, similar incidents happen in real life. Based on a survey of 31 countries, the cyber-security firm SOPHOS reported that 66% of healthcare organisations were hit by ransomware attacks in 2022, an increase of 94% compared to 2020.[2] Cyber-attacks have led to delayed chemotherapy,[3] the diversion of ambulances,[4] the failure of heart rate monitors,[5] and the breakdown of an intensive care unit,[6] causing the deaths of a baby in the US and a woman in Germany. Cyber-attacks are no longer just a threat to our data or money; they have become a threat to our health and lives.

The role of AI in this context is ambivalent. On the one hand, AI is urgently needed as a defensive tool. The sheer volume and variety of attacks, as well as velocity at which attacks now occur due to advances in the Internet of Things, cloud technology, and 5G, mean we require more effective protection. As a result, AI's capacity to process huge amounts of data very quickly has already and will increasingly become an invaluable tool in cyber-security. On the other hand, AI compounds the severity of threats because it too can be used in malicious ways. Generative adversarial networks (GAN) create deep fakes or forgeries of medical images, for example, by either adding or removing signs of lung cancer in CT scans. AI systems like DeepLocker take advantage of the fact that many AI systems are black boxes; since we cannot understand them, we cannot detect malicious intent. An AI system like PassGan breaks passwords with unprecedented success.

Let us imagine that Paul Pugsley and his team faced such scenarios for real: they *are*, after all, very real. In this chapter, I shall discuss what the key principles in biomedical ethics, such as autonomy, beneficence, non-maleficence, and justice, imply for countering cyber-risks in medicine.

Autonomy

Respect for patient autonomy requires, among other things, that a physician disclose relevant information and potential benefits and risks to the patient before the patient undergoes a medical procedure. Paul and his team might wonder what they should tell people in the future about the various cyber-threats that did not previously exist.

Cyber-threats are on the rise and are potentially very harmful. Yet, they are not *medical* risks and fall outside the scope of traditional disclosure requirements. Medical professionals are medical experts, and their duties are restricted to medical facts and medical risks. They are not IT experts or detectives who can inform people about what criminal hackers might do to them.

However, we shouldn't brush away the obligation to disclose cyber-risks so quickly. After all, with the use of AI-powered and digital medicine increasing, cyber-risks will be inseparable from medical care in the future.

To approach the matter, let's draw a distinction between safety and security. 'Safety' means the reliability and proper functioning of cyber-systems. Risks for safety are risks of malfunction, due either to technical glitches or benign (inadvertent) human error. On the other hand, 'security' means resilience to the deliberate, and normally malicious, interference by others. Risks for security are, then, risks of successful cyber-attacks. Ensuring safety requires protection against *accidental* harm, whereas ensuring security requires protection against *intended* harm.

Risks of cyber-*safety* require disclosure. In the future of medicine, digital and AI-based therapies will include, among other things, different implants such as cardiac pacemakers, insulin pumps, biosensors, and cochlear implants which are wirelessly networked and linked to other systems in order to 'monitor functionality, set parameters, exchange data or install software updates'.[7] Each of these implants will have a distinctive cyber-safety profile and these profiles could be highly significant for people, depending on their private and professional activities. If so, patients can legitimately demand to be informed about risks of cyber-safety. Without knowing the risks, they could not make informed decisions about their treatment options.

In addition, these safety risks meet a key criterion for disclosure. Lawyers and philosophers alike have argued that risks require disclosure when they are 'inherent' in a medical procedure where 'inherent' means that the 'risk is one which exists in, and is inseparable from, the procedure itself'.[8] Safety risks in digital medicine fulfil this condition because certain medical procedures, based on AI and other digital technologies, cannot be separated from them. Thus, as a matter of consistency, safety risks require disclosure.

There is an important implication at this point. The fact that these risks of safety are not risks of biological or pharmacological side-effects from certain drugs or treatments does not preclude them from being relevant information. Therefore, future disclosure needs to widen its perspective, by going beyond *medical* information, narrowly conceived.

But what about risks of *security*, rather than safety? There is an overlap with the previous arguments. Different medical procedures will not only have a distinctive cyber-*safety* profile, but also a distinctive cyber-*security* profile. Different AI-based and digital technologies are subject to different attacks, and sometimes AI systems introduce completely new vulnerabilities. Some deep neural networks in medical diagnostics are now vulnerable to so-called 'input attacks', which manipulate input data in ways that are often impossible to detect but which will frustrate an AI system in its diagnosis.

And it is only these AI systems, and not traditional computers, that are vulnerable to such attacks.

What is more, depending on a person's situation, some of these risks could be especially relevant, increasing *personal* risk. It's not too fanciful to imagine that in the future, the exploitation of AI systems could be used to attack specific individuals. In 2014, it was reported that former US Vice President Dick Cheney's pacemaker was vulnerable to attack and had to be deactivated for that reason.[9] Moreover, risks may also depend on the circumstances: for instance, cyber-attacks are more likely to occur in the midst of a pandemic because whenever there are crises, distracting health organisations, hackers face lower resilience in their targets. The World Economic Forum reported a 50.1% increase in cyber-attacks during the COVID-19 pandemic and identified 30,000 COVID-19 related cyber-attacks between 31 December 2019 and 14 April 2020. Among the most concerning attacks was the one on the Brno University Hospital in the Czech Republic that impeded emergency medical care and required the postponement of urgent surgery.[10] Thus, at least when there is increased personal or circumstantial risk, there is legitimate demand for information about the cyber-security profile of one's healthcare.

But as mentioned before, it is usually considered a condition of the requirement for disclosure that a risk is 'inherent' in a medical procedure. So, are cyber-security risks 'inherent' in this way? Scholars sometimes conflate two ideas in the notion of 'inherent', namely that something is 'internal' to a medical procedure or that something is 'inseparable' from a medical procedure. Traditional medical risks as well as cyber-safety risks are both: they are internal to and inseparable from medical procedures.

Security risks are not about internal failings of a system. They are instead about external attacks. Yet, the risks of such attacks are inseparable from certain AI-based medical procedures in the sense that it is not possible to employ those procedures without also incurring these risks, in particular when AI systems introduce new, hitherto non-existent vulnerabilities, such as the risk of 'input attacks'. So, there is at least one important aspect of 'inherence' that applies to certain risks of cyber-security too.

I therefore recommend a nuanced conclusion. Risks of cyber-security should not always require disclosure. After all, some cyber-attacks may not *at this point in time* be feasible, so the likelihood of such attacks could fall below a relevant threshold. Yet, there are no principled reasons against disclosure either. Individual patients can have a legitimate demand for information, at least in cases of increased personal and circumstantial risk and also because cyber-security is in at least one sense 'inherent' in the future of digital medicine. The principle of respect for autonomy requires updating clinical guidelines for the future so that they not only include the requirement to disclose information about cyber-*safety* but also, at least in some situations, cyber-*security*.

Beneficence

Let us grant that there is sometimes a requirement to disclose *more* information to future patients. Perhaps there is also a requirement to disclose *less*. Restricting the disclosure of information could be demanded by another key principle in medical ethics, the principle of beneficence, which obliges physicians to protect and promote patients' well-being. Traditionally, the principle of beneficence has been used to justify restricted disclosure in two main contexts, although this matter has remained controversial:

(i) In the context of the so-called 'therapeutic privilege', which states that a physician may legitimately withhold informa-tion based on a sound medical judgement that divulging the information would potentially harm a depressed, emotion-ally drained, or unstable patient.'[11]

(ii) In the context of placebos, when information about the pla-cebo's pharmacological or biomedical inertness is withheld to create health-conducive optimism in a patient.

In both contexts, withholding information is motivated by *benefi-cence*, since it purports to protect and promote the patient's well-being, or at least to avert further harm. Moreover, in both contexts, withholding information is *paternalistic* because the means of promoting well-being requires ignoring, overruling, or at least bypassing a patient's own will or judgement.

Now suppose that disclosing the risk of a cyber-attack causes such anxiety in a patient that they become unable to consider their options rationally, or highly overestimate the significance of the risk. If so, the considerations that purport to justify a therapeutic privilege may equally apply to not disclosing risks of cyber-attacks. The norms for the so-called therapeutic privilege might also apply to disclosure in the context of cyber-security.

With the increasing use of computer systems, there is another context in which disclosure could be restricted, namely:

(iii)   when information about certain systems could, if made public, make it easier for hackers to launch an attack. Not disclosing such information, for example where certain information is stored, how it is processed, the code of an AI system, can therefore be an important precautionary measure to increase patients' security.

Here, too, withholding information is motivated by beneficence, since it purports to protect and promote well-being, in this case through ensuring cyber-security. It is also paternalistic, insofar as physicians would withhold information even if patients requested it. Withholding information would be based on a negative judgement about the patients' ability to keep the data confidential enough to guarantee security.

There are two important differences between this context and the context of therapeutic privilege: first, withholding information to ensure cyber-security is no longer based on the prediction that the *disclosure itself* will harm the patient, for example by causing anxiety or irrationality, but rather that public transparency could encourage third-party attacks; second, the risk of a cyber-attack itself will still be disclosed. Only certain technical information about the digital or AI-based tools will be withheld from the patient.

Such non-disclosure would be part of what is sometimes discussed under the heading of 'security by obscurity', meaning the attempt to increase security by concealing or hiding information. 'Security by obscurity' is contentious in its specific applications, its advantages over full transparency, and its role in an overall security framework. Yet, several scholars have suggested that it can make an important contribution in certain settings,[12] of which the future of healthcare could be one. In any case, future medical ethics will surely have to enter this debate.

What is most noteworthy is that, from the perspective of medical ethics, the traditional criticism against non-disclosure in the contexts of the therapeutic privilege and the use of placebos does not equally apply to the novel area of 'security by obscurity'. The traditional criticism is that non-disclosure disrespects people as autonomous decision-makers, undermines trust in the patient–physician relationship, and can even backfire when people suffer harm through informational deficits. These criticisms do not apply to the novel area of 'security by obscurity' in just the same way, mainly due to two differences. First, with 'security by obscurity', medical professionals can employ what we may term second-order transparency: they can disclose to patients that there is certain information that they will not share with them, even if requested, since doing so would increase cyber-risks. By contrast, second-order transparency is not feasible elsewhere as it would contravene the therapeutic privilege and undermine the placebo effect. Thus, insofar as greater transparency is conducive to respecting patient autonomy, non-disclosure in this novel field becomes less problematic. Second, non-disclosure in the area of 'security by obscurity' is not *as* paternalistic as it is in the contexts of the therapeutic privilege and the use of placebos. This is because not disclosing cyber-related information is necessary to protect not only one patient but potentially a larger group of other patients who are also subject to similar types of digital and AI-based healthcare. For this reason, medical professionals could be under an additional obligation of beneficence, owed to other patients, that they do not increase the risk of harm to them by disclosing sensitive information to others. And insofar as the degree of paternalism affects the force of the criticism, non-disclosure may become less problematic for this reason too.

Thus, with the increasing use of computer systems in medicine, we may find not only another context in which the principle of beneficence proposes paternalistic reasons to withhold information, but also an area where withholding information is better justified. For this reason, future guidelines need not only ask for *more* disclosure, based on the principle of autonomy, but also consider *less* disclosure, based on the principle of beneficence.

Non-maleficence and Justice

So far, we have only considered what to do *ahead of* cyber-attacks. But surely, we also need to think about what to do *during* a cyber-attack. There are different ways to respond to an attack, for example shutting down one's own system and hoping to recover it, trying to identify the attacker, or even pursuing an 'aggressive response'. Often, shutting down a system and recovering the data is not feasible, either because recovery is not possible or because it would take too long. Moreover, with sufficiently sophisticated attacks, identifying an attacker is often impossible. So what about counter-attacks? Could they be justified?

Medical professionals are bound by the Hippocratic Oath: above all, do no harm! Today, this is expressed in the principle of non-maleficence. How far does this principle extend? Clearly medical professionals and their wider team should not harm patients, but what about other people, especially those who pose a threat to patients? In the drill, Paul Pugsley's patient was attacked by someone; couldn't his medical team fight back, even if this harms the attacker?

In addition to the principle of non-maleficence, medical professionals are also bound by the principle of justice, which requires 'fair, equitable, and appropriate treatment in light of what is due or owed to affected individuals and groups'.[13] Typically, the principle of justice concerns the distribution of benefits and burdens. But so understood, justice also entitles people to forms of defence, including self-defence and other-defence. If someone attacks a person, they forfeit their right not to be harmed as part of a proportionate response. People who are attacked do not need to bear the cost of the attack if they, or others, can avert it by harming the attacker.

If we look at the two principles together, there will be occasions when medical teams are permitted to fight back. Justice sets limits to non-maleficence and allows for defence. What is more, averting harm to the patient by a counter-attack could also be required by the principle of beneficence that obliges physicians to protect their patients' well-being. If so, a countermeasure could even be obligatory, rather than merely permissible.

Cases of permissible and even obligatory counter-attacks will be those when the countermeasure is *necessary* to avert the harm, *proportionate* in comparison to the gravity of the threat, and *directed* against the attacker only. Take the attack on the intensive care unit in Düsseldorf in 2020, when hackers disabled the computer systems of the hospital while a woman was in a critical condition. Unfortunately, the medical team couldn't avert the attack and had to transfer the woman to another hospital. The transfer took so long that the woman died.[14] But suppose that IT experts in the hospital could have identified the attacker and averted the attack by a countermeasure, which would have destroyed the attacker's computer and erased all of their files without affecting other people. In this case, the countermeasure would have been *necessary*, certainly *proportionate*, and also *directed*, and therefore clearly justified.

In philosophy, this common-sense principle of defence is hardly novel. For example, it is regarded as clearly permissible to temporarily incapacitate someone who is about to attack another person with a knife. In this case, just as in my hypothetical version of the Düsseldorf situation, the defence would be necessary, proportionate, and directed, and therefore permissible. The only unusual aspect of the hypothetical Düsseldorf case is that it shows this defence-principle now also applies to *medical facilities* and permits, or even obliges, these facilities to fight back under certain circumstances.

But in the medical world, there are hardly ever clear-cut scenarios. To begin with, it will not often be clear whether a counter-attack will avert rather than escalate a threat, leading to punitive responses by the attacker and potentially further harm to patients.

Moreover, many attackers launch their attacks, especially so-called distributed denial of service attacks (abbreviated as DDOS) through innocent third-party machines to conceal their identity, so inevitably counter-attacks will harm innocent third parties too. Worse still, one would probably not know what kind of harm could be caused to these innocent parties. In some cases, it might only be a minor breach of privacy. In others, a counter-attack could cause severe harm. The philosopher Kenneth Himma illustrates the point in this way:

> Suppose, for example, that an attacker compromises machines on a university network linked to a university hospital. If hospital machines performing a life-saving function are linked to the network, an aggressive response against that network might result in a loss of human life.[15]

Given these challenges, medical facilities of the future need to develop new capabilities. They need to acquire the ability to counter an attack whenever the attacker alone can be targeted or, potentially, also, in cases of marginal harm to innocent third parties. Moreover, medical facilities need to acquire the ability to investigate attacks quickly and to cooperate with authorities as efficiently as possible. At the same time, however, they need to be prepared for unclear cases and to produce an ethics protocol, stating actionable advice on, among other things, how to deal with uncertainty surrounding the potential harm of a counter-attack and how to weigh the different rights and interests of the patients attacked and the innocent third parties affected. The principles of justice and non-maleficence can provide some guidance here, but they require us to think afresh about the challenges of cyber-security in medicine.

Conclusion

Cyber-risks are a growing concern in medicine and require us to update clinical ethics. My aim in this chapter has been to highlight the implication of four key principles in biomedical ethics for this new world. The principle of autonomy implies more transparency regarding the cyber-profile of medical procedures (introducing a novel category for disclosure), while the principle of beneficence may justify less transparency for the sake of increased cyber-security (introducing a novel category of restricted disclosure). Finally, the principles of non-maleficence and justice together permit counter-attacks under certain limited conditions and emphasise the need to develop protocols for situations of uncertainty. With these tools, we can update clinical ethics for the challenges of digital and AI-based medicine in the 21st century and make sure medical professionals like Paul Pugsley and his team are prepared for what's ahead.

Acknowledgements

I am very grateful to David Edmonds for his truly outstanding editorial support and very constructive comments on earlier versions of this chapter. I thank Claudia Negri Ribalta, Marius Lombard-Platet, and Thomas Douglas for their very helpful feedback. I thank too the Leverhulme Trust for their generous financial support (project ID: *ECF-2021-176*).

Notes

1. <https://www.theverge.com/2019/4/4/18293817/cybersecurity-hospitals-health-care-scan-simulation>.
2. <https://assets.sophos.com/X24WTUEQ/at/4wxp262kpf84t3bxf32wrctm/sophos-state-of-ransomware-healthcare-2022-wp.pdf>.
3. <https://www.route-fifty.com/public-safety/2022/05/ransomware-attacks-hospitals-put-patients-risk/367117/>.
4. <https://www.sandiegouniontribune.com/news/health/story/2021-05-05/state-regulator-watching-scripps-ransomware-attack-closely>.
5. <https://www.wsj.com/articles/ransomware-hackers-hospital-first-alleged-death-11633008116>.
6. <https://www.wired.co.uk/article/ransomware-hospital-death-germany>.
7. K. Weber and N. Kleine, 'Cybersecurity in Health Care', in M. Christen, B. Gordijn, and M. Loi (eds), *The Ethics of Cybersecurity* (Springer, 2020), 146.
8. *Jones v. Papp* [1989] 782 S.W.2d 236 (Tex.App.).
9. <http://www.informationweek.com/healthcare/security-and-privacy/dhs-investigates-dozens-of-medical-device-cybersecurity-flaws-/d/d-id/1316882>.

10. H. S. Lallie, L. A. Shepherd, J. R. Nurse, A. Erola, G. Epiphaniou, C. Maple, and X. Bellekens, 'Cyber Security in the Age of Covid-19: A Timeline and Analysis of Cyber-crime and Cyber-attacks during the Pandemic', *Computers & Security* 105 (2021): 102248.

11. T. L. Beauchamp and J. F. Childress, *Principles of Biomedical Ethics* (8th edn) (Oxford University Press, 2019), 126.

12. J. C. Smith, 'Effective Security by Obscurity', in *arXiv preprint arXiv:2205.01547* (2022).

13. Beauchamp and Childress, *Principles of Biomedical Ethics*, 267–8.

14. <https://www.wired.co.uk/article/ransomware-hospital-death-germany>.

15. K. E. Himma, 'Ethical Issues Involving Computer Security: Hacking, Hacktivism, and Counterhacking', in K. E. Himma and H. T. Tavani (eds), *The Handbook of Information and Computer Ethics* (John Wiley & Sons, 2008), 211.

Further Reading

Markus Christen, Bert Gordijn, and Michele Loi, *The Ethics of Cybersecurity* (Springer Nature, 2020).

3

Risky Business

AI and the Future of Insurance

Jonathan Pugh

Filling in an insurance form can be a tedious yet morbid exercise. It typically involves responding to a deluge of questions about yourself, some of which appear to bear only the faintest relationship to the matter at hand—namely, establishing the likelihood that something bad is going to happen to you.

Whilst the risk of bad things happening is an immutable part of life, the way in which insurers calculate that risk is changing. As in so many areas, artificial intelligence is transforming the nature of the insurance industry; in addition to streamlining claims-processing, improving customer service with chatbots, and detecting fraud, AI is also increasingly being used in the underwriting and pricing process. It offers the prospect of harnessing big data to deliver detailed and accurate risk classification to inform the pricing of insurance premiums. As well as employing algorithms to identify patterns that might evade a human actuary, insurers may also be able to draw on a larger pool of information, using data 'hoovered' from the consumer's online life, their wearable healthcare technology, and other smart devices incorporated into the 'Internet of Things'. Ultimately, AI promises to deliver highly individualised insurance pricing. This lies beyond the capabilities of today's human actuaries, who typically use information to place groups of customers into broader risk pools.

This might make today's lengthy form-filling exercise appear more palatable. Those who champion privacy may prefer to exert greater control over the information shared with insurers. Furthermore, you might be concerned about whether consumers are really giving consent to companies when they 'click' their data away online. Whilst these are important concerns, I shall set them aside here, merely noting that some consumers appear highly willing to proactively volunteer data from their mobile health technologies (e.g. Fitbits) to obtain insurance discounts.[1] Instead, I shall discuss a broader range of moral concerns grounded in considerations of justice. I will argue that these considerations support increasing the oversight of AI in insurance to limit the extent to which it can be used to determine insurance pricing. To begin, it will be helpful to reflect on the nature of insurance, and how different conceptions of fairness interact with the industry.

Types of Insurance and Types of Fairness

Fundamentally, the role of insurance is to protect individuals from the financial consequences of catastrophic risk events, whose effects would otherwise be individually unaffordable. Insurers offer this protection by spreading the risks and costs of these events, by collecting financial contributions from many people to pay for risk events that befall anyone within that group. In this way, solidarity lies at the heart of the very idea of insurance; the worst could happen to anybody, so we should all agree to help each other.

There are different ways in which an insurance scheme can be funded. First, governments might run *social insurance* schemes; here, citizens will typically be required to contribute (e.g. via taxation) to the government's providing a certain level of protection to all members of society. That may include, for instance, access to healthcare or unemployment benefits. Although different participants in social insurance schemes may contribute different amounts (e.g. the wealthy may contribute more than the poor), all citizens will receive a similar level of cover should they need it.

Social insurance can be understood as an application of a *sufficientarian* approach to distributive justice. Sufficientarians claim that a central goal of justice is to ensure that people 'have enough'; justice is about ensuring individuals receive a level of some good above a minimum threshold of sufficiency. In this vein, social insurance can be understood as a kind of 'safety net', preventing people from falling below a sufficiently good standard of living. A key question of justice regarding such schemes is how governments should determine where to set this threshold of sufficiency. In doing so, they must manage trade-offs between the moral reasons to minimise the imposition of mandatory financial contributions upon citizens, and the reasons to ensure that the social safety net is 'high' enough to catch those who need it before they fall below a reasonably good quality of life.

Social schemes can be contrasted with private *mutualistic insurance* schemes; to somewhat abuse the above metaphor, mutualistic schemes allow individuals to buy a 'higher' safety net than that which the state arranges via social insurance. Unlike social schemes, both the level of insurance an individual can access, and the amount they have to pay for it on a mutualistic model, depends on their risk profile; broadly, the higher the risk of the insurance company having to pay out, the higher the premium the customer has to pay for it. This, for example, is why smokers typically pay higher premiums for life insurance than non-smokers. When the magnitude or probability of a risk event is uncertain, insurers may even deem some risks to be uninsurable.

Smokers and non-smokers are thus subject to differential treatment in the context of life insurance. Is this fair? One argument in favour of claiming that it is could appeal to the 'principle of equity':[2]

Principle of Equity: Policy-holders should make financial contributions that are commensurate to their degree of risk.

This principle can be understood to ground an actuarial conception of fairness, whereby fairness is understood to require 'paying your dues'; on this understanding, it is unfair for a high-risk customer to be charged a low premium, or a low-risk customer to be charged a high premium. It would also be unfair to charge high-risk and low-risk customers the same premium, as this would effectively amount to the low-risk group subsidising the high-risk group (so-called 'cross-subsidisation').

The actuarial understanding of fairness can be traced back to a broadly libertarian approach to justice and ideals of 'self-ownership'; the idea that people should be free to reap the rewards of their natural assets. To abide by this actuarial conception of fairness, insurance companies need to make reasonably accurate risk assessments. Without accurate assessments, companies face the prospect of adverse selection, a phenomenon that can threaten the viability of the insurance market. Adverse selection occurs when insurers cannot adequately distinguish high-risk from low-risk customers; as a result, they have to charge high average premiums that are unattractive for clients who believe themselves to be low risk. For example, suppose a 40-year-old customer has received a recent unexpected medical diagnosis of a terminal condition that will almost certainly kill him within the next 12 months. This customer would benefit considerably from a life insurance policy that would help his young family cover the financial costs associated with his unexpected death. Now, suppose that he was not obliged to disclose this diagnosis to an insurance company when buying insurance; if he withheld this information, he would be able to obtain a policy for the low prices that companies reserve for other 40-year-old clients believed to be at low risk of death. However, imagine that this happened at a larger scale—a large number of people receive this diagnosis, and none disclose it when buying life insurance. As more and more deaths occur in the apparently 'low-risk' group of customers, the company would have to increase premiums for everyone in this 'low-risk' pool to cover the pay-outs. But this will serve to disincentivise the genuinely low-risk clients (i.e. those younger adults without a terminal diagnosis) from purchasing policies at the now higher rates. If enough high-risk customers are inaccurately placed into a low-risk pool, this can lead to escalating costs across the sector, and even market failure.

Therefore, actuarial fairness speaks in favour of increasing the accuracy of risk prediction. However, the concept of actuarial fairness does not exhaust the ideals of justice relevant to insurance. For instance, consider that it is not actuarially unfair if insurance becomes too expensive for the highest-risk individuals to access; to use the previous example, it would not necessarily be unfair in this sense for insurance companies to only provide life insurance at an astronomical price to an individual who disclosed their diagnosis, or even for them to refuse coverage to that individual if their risk of premature death amounted to a certainty. Moreover, cross-subsidisation also runs contrary to the concept of actuarial fairness. Yet, some might claim that *it is fair* for low-risk people to subsidise high-risk customers, if this amounts to a redistribution of unearned benefits and burdens. Indeed, notice that the principle of equity does not capture the idea that degrees of risk will often not be fairly distributed across individuals in the first place. Sometimes, risk is a matter of pure luck; for example, you may be born with a genetic predisposition to certain diseases.

Accordingly, actuarial fairness in insurance plausibly needs to be supplemented with further principles of *moral* fairness. These principles could be based on various theories of justice, or deontological constraints against discrimination. Luck egalitarianism is one frequently invoked theory in this context; a luck egalitarian principle would aim to neutralise the effects of luck on an individual's risk profile, and claim that people should not be charged more for risk factors that they cannot control. To illustrate, on this view it might be fair to charge smokers higher premiums for life insurance if they voluntarily choose to smoke, but it would not be fair to charge men a higher premium for car insurance, even if they have a statistically much higher chance of being involved in an accident.

These brief remarks are enough to show that actuarial fairness and moral fairness can pull in different directions, and this can imply trade-offs between the accuracy of risk profiling (required for actuarial fairness) and conceptions of moral fairness. Consider, for example, the 2011 EU court ruling that stipulates that the use of gender information in insurance must not result in differences in individuals' premiums and benefits (at least for insurance that does not cover gender-specific conditions).[3] Critics of this move claimed that it failed to acknowledge the predictive value of gender information in many insurance domains. However, the predictive validity of gender information was not the issue at stake in this debate;

rather the issue was one of how to manage the trade-off between actuarial fairness and a particular conception of moral fairness.

Finally, it might be argued that it is *procedurally* unjust for insurers to use information in underwriting and pricing when its predictive value cannot be adequately explained to consumers. The idea of procedural injustice here is that there can be unfair ways of making decisions about pricing, even if the resulting price is itself fair. To illustrate, in some cases the relationship between a given risk factor and a risk event is quite straightforward. For example, we have a good understanding of why smoking increases an individual's mortality risk. However, in other cases, the relationship is less straightforward; consider the fact that some insurers use consumer credit scores in costing insurance policies.[4] One reason that this practice is controversial is that, even assuming there is a statistically valid association between one's credit score and one's risk of a catastrophic event (such as premature death), it is difficult for consumers to intuitively grasp the relationship between the two. Due to the lack of transparency involved, and the manner in which it may prevent consumers from seeking redress, using such information to set insurance premiums without further explanation can thus be understood to amount to a procedural injustice—even if the price accurately reflects the customer's risk profile. Furthermore, in the absence of an explanation, consumers may harbour the belief that this relationship between credit scores and catastrophic events is merely a 'spurious correlation'—a relationship between two variables that appears to be causal, but which is not, and may actually be explained by a further unrelated and perhaps unknown 'lurking' variable.[5] Such suspicions can undermine trust in the fairness of underwriting and pricing. For example, there is a strong (around 95%) but clearly spurious correlation between the annual number of deaths involving fishing boats, and the marriage rate in Kentucky[6]—however, this information is hardly a sound basis for trustworthy life insurance pricing.

In view of the above, we can identify various justice-based concerns regarding the use of AI in insurance. One relevant issue is familiar from other domains of AI—the problem of opacity.

Suppose you were denied insurance coverage following analysis from an 'AI actuary'; the AI actuary has identified a host of highly predictive, but surprising risk factors from data recovered from your online activity and monitoring devices. Suppose that it has even performed complex analyses to determine that these factors are not spurious correlations. However, due to the black box nature of machine-learning, the details of the relationship between the newly identified risk factor and the risk event may remain unintelligible. You might plausibly complain that being denied insurance on this basis constituted a form of injustice; your complaint here would be that the industry has failed to meet a plausible requirement of procedural justice in this case.

The opacity issue might admit of a technological solution—but AI arguably raises other issues of justice for which technological solutions are less obvious. As AI achieves the goal of making risk prediction more accurate and personalised, this raises the possibility that the highest-risk individuals will be priced out of obtaining insurance, generating an insurance 'underclass'. Further, if the lowest-risk individuals develop a better understanding of their own personal risk profiles (informed, for example, by wearable health technologies etc.) this may reduce their motivation to purchase insurance. Accordingly, although increasing personalisation in the industry may promote actuarial fairness, there is a legitimate concern in the industry that it may also serve to divide society into those who don't need insurance, and those who are too expensive to insure.[7] Personalisation thus threatens the solidarity that lies at the heart of insurance, and the ability of the most vulnerable to access insurance through a mutualistic scheme.

The extent of this threat is ultimately an empirical and economic question; notably, there are also ways in which AI could extend access to insurance over the long term. For instance, if AI analysis can reduce the uncertainty pertaining to the probability of certain risk events, then this will reduce the number of uninsurable risks. Nonetheless, the prospect of an insurance underclass in the AI age is one that has been recognised within the industry,[8] and it is therefore worth considering the ethical values at stake in potential solutions.

One solution would be to prohibit the use of AI systems that can deliver highly personalised risk profiles. However, in addition to doubts about enforceability, there are two ethical concerns about a prohibition approach. First, the use of AI in underwriting can lead to other important moral goods beyond actuarial fairness. Information garnered from industry grade AI risk prediction could enable consumers to mitigate or prevent harms; indeed, some believe that the use of AI might move the industry from a 'detect and repair' framework to a 'prevent and protect model'.[9] For example, if insurers realise that individuals are gradually increasing their risk of cardiovascular disease due to physical inactivity (as monitored by the customer's wearable devices), they could inform the customer of their increasing risk, and even seek to incentivise physical activity for that consumer. Whilst this somewhat paternalistic form of intervention will naturally be welcomed by broadly utilitarian approaches to the ethics of insurance, it can also have downstream effects on fairness, insofar as poor health disproportionately affects the least advantaged members of society.[10]

Second, we should be cautious about restricting the ability of insurers to access predictive information that remains available to clients. As detailed above, such informational asymmetries can provide fertile ground for adverse selection. As the power of consumer smart technology advances, consumers themselves may be able to develop a more sophisticated understanding of their own personal risk profiles.

Perhaps the perils of increasing personalisation in insurance might suggest that we should abandon mutualistic insurance models for social models, or at least radically expand the scope of the latter. This would be a neat solution to the issue of adverse selection, as such models are not threatened by this phenomenon— individual insurance pricing and coverage are not determined by risk on such an approach. However, abandoning mutualistic insurance schemes may not be politically feasible, and would run contrary to the broadly libertarian conception of justice underlying the notion of actuarial fairness. The debate about whether to provide certain kinds of insurance via a social or mutualistic scheme can be understood as a manifestation of a fundamental political conflict between fairness and liberty. Considerations of fairness (sourced from a range of theories of justice) can lend support to social insurance models. However, abandoning mutualistic models in favour of social insurance would plausibly require significant increases in mandatory financial contributions from citizens, and prevent them paying into a private mutualistic scheme. This would therefore involve a considerable cost to liberty. Of course, when the provision of insurance constitutes a basic need, reasons of moral fairness may plausibly override considerations of liberty. Conversely, if the insurance in question goes beyond safeguarding basic needs in a society, it is less clear that considerations of fairness can be invoked to justify the infringement of liberty involved in providing that insurance via a social scheme.

So, is there a way of reaping the potential benefits of AI in mutualistic insurance, whilst forging a balance between fairness and liberty that adequately protects the most vulnerable? Two existing policy mechanisms in the insurance industry might help. One is to allow a market for mutualistic insurance in which risk classification is informed by AI, but where the government regulates the industry to ensure that the highest-risk individuals can still access affordable insurance. Consider the approach taken to flood insurance in the UK, where the government (in collaboration with the industry) introduced the Flood Re scheme. Under this scheme UK household insurers are required to pay into a fund that can be used to pay off claims for the highest-risk homes.[11] Notice that on this approach it is the insurance companies, rather than consumers, whose liberty is infringed to safeguard fairness.

An alternative solution is to limit the degree of coverage that certain methods of risk classification can be used to price. Again, this has some precedent in the industry in the context of genetic testing. Despite their predictive power, many countries prohibit the use of predictive genetic test results in insurance underwriting, due to concerns about genetic privacy, discrimination, and fairness. However, in the UK, insurers are permitted to make some limited use of these results. The Association of British Insurers adopts a code of practice that permits companies to ask for and take account of predictive genetic test results for specified diseases when underwriting Life Insurance, Critical Illness Insurance, and Income Protection Insurance. Crucially though, the code of practice stipulates that they may only do so for policies exceeding stipulated financial thresholds (e.g. £500,000 per person for life insurance).[12]

In this vein, one could similarly limit the use of highly personalised AI risk prediction to the pricing of policies that exceed a stipulated financial threshold. This would be another way of balancing the competing demands of moral fairness, actuarial fairness, and liberty, in the face of an increasingly personalised approach to risk classification. The ethical challenge of this approach is deciding where to set the relevant financial thresholds. The aim should be for considerations of actuarial fairness to dominate only for the pricing of policies that are unnecessary for the demands of moral fairness.

This challenge is a permutation of a much broader question that the use of AI in insurance is forcing us to confront: just how much protection should a just society be expected to provide for its citizens, and how much freedom should people have to benefit from their natural assets? Until now, uncertainty about our own risk of catastrophe has meant that our individual self-interest has aligned with moral reasons of solidarity to contribute collectively to mutualistic insurance schemes—bad things could happen to you, but they could also happen to me. If we are uncertain about who will be the victim of a tragedy, it is in our collective interests to pay to look after each other, even when the state does not mandate it. By reducing this uncertainty, AI is moving the practical recommendations of morality and self-interest in this sphere further apart.

If we are to safeguard the solidarity that forms the very basis of insurance, and ideals of justice in this sphere, then we need regulation. Sometimes the state itself is obliged to provide insurance as a basic need. But where this is not the case, we should limit the power to which AI can determine the price that people pay to access this crucial good.

Notes

1. Andrew Boyd, 'Could Your Fitbit Data Be Used to Deny You Health Insurance?', *The Conversation* (2017), <http://theconversation.com/could-your-fitbit-data-be-used-to-deny-you-health-insurance-72565>.
2. Jonathan Pugh, 'Genetic Information, Insurance and a Pluralistic Approach to Justice', *Journal of Medical Ethics* 47/7 (1 July 2021): 473–9.
3. 'Insurance Gender Ruling and You', *BBC News*, Business, 1 March 2011, <https://www.bbc.com/news/business-12608777>.
4. 'Credit-Based Insurance Scores', accessed 8 December 2022, <https://content.naic.org/cipr-topics/credit-based-insurance-scores>.
5. 'Spurious Correlations', accessed 8 December 2022, <http://tylervigen.com/spurious-correlations>.
6. 'Spurious Correlations'.
7. 'Aviva GI CEO on Claims Inflation, Stupid Drivers and Commercial Insurance', accessed 8 December 2022, <https://www.nsinsurance.com/analysis/aviva-colm-holmes-claims-inflation/>.
8. Marc Ambasna-Jones, 'The Smart Home and a Data Underclass', *The Guardian*, Media Network, 3 August 2015, <https://www.theguardian.com/media-network/2015/aug/03/smart-home-data-underclass-internet-of-things>.
9. 'Insurance 2030—The Impact of AI on the Future of Insurance', accessed 8 November 2022, <https://www.mckinsey.com/industries/financial-services/our-insights/insurance-2030-the-impact-of-ai-on-the-future-of-insurance>.
10. George Loewenstein, David A. Asch, and Kevin G. Volpp, 'Behavioral Economics Holds Potential to Deliver Better Results for Patients, Insurers, and Employers', *Health Affairs (Project Hope)* 32/7 (July 2013): 1244–50, <https://doi.org/10.1377/hlthaff.2012.1163>.
11. 'How Flood Re Works', Flood Re, accessed 7 December 2022, <https://www.floodre.co.uk/how-flood-re-works/>.
12. Association of British Insurers and HM Government, *Code on Genetic Testing and Insurance* (HM Government, 2018 [updated 2022]), <https://www.gov.uk/government/publications/code-on-genetic-testing-and-insurance>.

Further Reading

Kevin Glaser, *Inside the Insurance Industry* (Right Side Creations, 2014).

4

AI and Discriminatory Intent

Binesh Hass

Introduction

Suppose you know that a particular area of town is a bad place to be after a certain hour. 'Avoid after dark, it's known for armed mugging.' Your fears are confirmed by the relevant data, which shows that muggings are, in fact, frequent after dark. Suppose the data also shows that the people doing the mugging are predominantly young, male, and 'ethnic' ('YME', for short). That is data about a group of people, that is, muggers.

One evening, you find that you need to traverse that part of town because you missed your bus. In the distance, walking towards you, you see a YME—'Uh oh,' you say to yourself as you think of ways to avoid him. In that moment, you have taken data about a group and applied it to a single person, and then formed an intention to do something. Is that justifiable?

One way to think about the question is by weighting the potential and actual harms involved. On one hand, there is the potential harm, sometimes called a 'doxastic harm', involved in a belief about someone that in some relevant way diminishes their standing in your mind as an autonomous individual with a moral claim not to be treated as a mere statistic. On the other hand, there is the potential harm associated with being mugged. On balance, you conclude that the offence you could give by racially stereotyping the person approaching you is less grave than what might follow from being mugged. 'Better to seem a racist than risk my life,' you think as you cross the road and quicken your pace.

The story I have just told is commonplace. One philosopher, who retold a version of it at a recent seminar, recalled having consciously decided not to act upon a racial stereotype as they, some evening after a talk they had given, approached a group of YMEs in a rough part of Boston. They were mugged at gunpoint (but were otherwise unharmed).

The story is also an analogue to a seemingly intractable problem about the role of AI in criminal law, particularly in the context of courts increasingly relying on algorithmic risk-assessment instruments to inform sentencing decisions. Should we give Amir a harsher sentence because of certain demographic features of his which an AI-based instrument uses to predict that he is more dangerous than Jeff?

Discrimination

AI-based risk assessment instruments in sentencing rely on machine learning algorithms which analyse features of a convicted defendant's identity for the purpose of predicting recidivism. These predictions are then used alongside other considerations to determine the length and terms of a sentence (e.g. the more likely you are to reoffend, the harsher your sentence). It is essential to these predictive systems that they match a feature of a defendant to a pattern in a dataset that suggests that others who share that feature have been more likely to reoffend (or break parole, etc.). That is similar to the calculation a person makes when they find themselves in a part of town where YMEs are disproportionately represented amongst muggers and are wondering whether the approaching YME person is trouble.

The predicament, of course, doesn't occur in a social vacuum. It has long been established that visible minorities are more likely to face arrest,[1] be charged,[2] end up behind bars for the same offences,[3] and receive significantly lengthier sentences than similarly situated white offenders.[4] Datasets on police contact, trial, and sentencing are therefore likely to reflect these realities.[5] Accordingly, predictions of a defendant's recidivism sourced from these datasets will be problematic to the extent that they rely on the over-representation of

visible minorities (and under-representation of white individuals)—
'bias in, bias out,' as one study notably put it.[6] Racial disparities in
criminal justice systems are infamously widespread and entrenched,
and their complex causes are well studied.[7]

To address these disparities in the context of AI-based risk assessment systems, the law typically relies on discrimination statutes,
which distinguish between direct and indirect discrimination. The
former occurs when someone treats you unfavourably because of a
legally recognised 'protected characteristic'.[8] The textbook examples
of such characteristics are race and sex, but throughout the common law the list is more expansive and varies (e.g. in Canada, but
not the UK, genetic characteristics are afforded federal statutory
protection,[9] whereas in the UK, but not in Canada, certain of one's
non-religious beliefs are protected under statute).[10] A judge who
hands out lengthier sentences to black individuals because they are
black, for instance, is directly discriminating against blacks. On the
other hand, indirect discrimination, sometimes called 'disparate
impact', occurs when a prima facie neutral policy unfavourably and
disproportionately impacts those with a protected characteristic.
For example, a sentencing instrument might count educational
achievement as a mitigating factor in determining the length and
terms of a sentence; but in contexts where certain racial groups
have been deprived of equal access to education, that might constitute indirect discrimination. Legal systems throughout the common law prohibit both direct and indirect discrimination, but are
especially strict with the former.

Concerns about the use of AI-based risk assessment instruments
typically centre on indirect rather than direct discrimination. It is
easy enough to see why. If an AI-based instrument is legally prohibited from relying upon a protected characteristic such as race in
providing an assessment of a convicted defendant's risk of reoffend-
ing, then one might think that the only remaining way for such a
model to discriminate against race would be to find proxies for it,
such as by relying on postal or zip codes.[11]

Accordingly, the prevailing view in the literature has been that if
the AI-based instrument results in discrimination, it is going to be

of the indirect variety.[12] That could be an unwelcome conclusion if you find yourself on the receiving end of discriminatory policy, for indirect discrimination is 'almost always justifiable' in adjudicative settings.[13] So long as a discriminatory policy can be shown to be necessary and proportionate to securing an important enough end, then it is to that extent legally justified. In the context of a court trying to sentence a violent offender with a high risk of recidivism, the justificatory bar is likely to be satisfied.[14] Direct discrimination, in contrast, is almost always unjustifiable in adjudicative settings. That contrast in justifiability, in turn, prompts a neglected but important question in relation to the discriminatory potential of AI-based predictive instruments—the question, that is, of whether such instruments can be *directly* discriminatory.

Intent

That puzzle, in some jurisdictions, will turn on the role of intent. In the US, to show that a policy is directly discriminatory, you need to establish that the intention (or motive) of the policy was, in fact, to discriminate. That is obviously a high if not altogether impracticable bar in AI, since AI-based algorithms are incapable of intention. That, in any case, is the standard view in the legal literature: intention is a feature of human cognition, AI-based instruments cannot be said to have intent, and so such instruments cannot be said to be in the business of direct discrimination even when they are finding very good proxies for protected characteristics. 'Like trying to force a square peg into a round hole' is how one set of lawyers characterised the insistence on intent for establishing discriminatory purpose in the context of AI-based instruments.[15]

That line of thought has seemed so persuasive that it has led some to argue for intent to be replaced by an effects-based framework in evidencing discrimination. There are, indeed, good reasons for legal systems to avoid conceptualising intent as a subjective mental state. As one UK judge suggested, we should want a way of determining discrimination questions that avoids 'complicated questions relating to concepts such as intention, motive, reason or purpose, and the danger of confusion arising from the misuse of those elusive terms'.[16] That pragmatic desideratum is made partly possible by what is called the *but-for* test. The test provides a causal explanation for why things go as they do: but for the fact that you have some property *p* (e.g. pregnancy), things would have been otherwise (e.g. you would have been hired).

One advantage of a causal explanation is that it doesn't care for the reasoning behind a particular policy. Consider one of the UK's landmark discrimination cases, *James v. Eastleigh Borough Council*. In that case, the House of Lords had to decide whether a public authority, in providing free swimming pool access to pensioners, was directly discriminating against men given that the 'pensionable age' of men was higher than that of women (when Mr and Mrs James used the pool, Mr James had to pay 75p and Mrs James didn't). No one involved in the case doubted that Mr James's sex did *not* feature in the reasoning of the public authority. Eastleigh Borough Council was strictly concerned with age in providing free access. In spite of that, the court found that the Council had *directly* discriminated on the basis of Mr James's sex, since if it were not for his sex, he would not have been adversely affected by the Council's policy on who gets free access. The court held that, in intentionally relying on a factor (pensionable age) that entailed differential treatment for men and women, the Council intended to discriminate on the basis of sex (as it was then recognised). If you intended to A, and A entailed B, that's sufficient to establish that you intended B—or so the court argued. The main idea, in any case, is that intention is thought to be ascertainable through causal analysis (rather than through an investigation of mental states).

Now, a further benefit of the effects-based route, as Tarun Khaitan fruitfully explains, is that it can easily accommodate intersectional discrimination.[17] Take the case of the AI-based instrument which, in building an identity for profiling, settles on two or more features: the YME person is not profiled because of their youth, maleness, or ethnicity but because of a confluence of those features. But for the aggregate of these protected characteristics, there would be no profiling. Here, too, the road to establishing discrimination is free of the usual contortions involved in determining intent. Describing how the causal test operates in practice, Neil Gorsuch, in a recent US Supreme Court case, put it this way: we 'change one thing at a time'—or, indeed, one intersectional set of things at a time—'and see if the outcome changes. If it does, we have found a but-for cause.'[18]

There are, however, a few viable albeit neglected pathways for keeping intent in the story about direct discrimination and AI-based instruments. Typically, when we want to ascertain intent, we do so in ways that are specific to the domain of activity at hand. Criminal law, for instance, distinguishes between specific intent, where someone has a distinct end in mind (e.g. A is out to mug B), and general intent, where the end that motivates the unlawful act is less distinct (e.g. A is seeking to mug anyone in some general vicinity). The prosecution will need to show, beyond a reasonable doubt, that A had a particular mental state, that is, either specific or general intent. But obviously, unless A is forthcoming about their mental states, the prosecution is going to have to draw its own conclusions on the basis of A's actions (e.g. did A write something down, draw up a map, rehearse the crime?) and those actions need to strike A's peers in the jury as being revealing of A's intentions.

All of that is possible without getting into A's head. We just need to look at A's acts and ask the jury to decide if it's beyond reasonable doubt that those acts are revealing of the relevant category of intent. A similar doctrine of 'objective intent' could be brought to bear on AI-based risk assessment instruments. For, in addition to the ability to *act*, they can also *omit*—for example, by failing to 'correct' a data-set that is reasonably expected to yield discriminatory outputs by explicitly debiasing it. Thus an AI instrument might fail to under-sample for over-represented groups when it could easily do so, for example, by randomly removing individuals from the over-represented group to approximate their representation in the relevant community-level or regional-level population. Or an AI instrument may fail to enforce strict fairness constraints, for example, by prohibiting use of variables, such as postal or zip codes, which can be proxies for protected characteristics in certain regions.

A failure to debias when bias is patently obvious is akin, if not tantamount, to the intent to let that bias prevail. When that bias is discriminatory, the intention is to let the discriminatory bias prevail. Provided that the bias targets a protected characteristic, what we have is nothing less than the intention to let the direct discrimination prevail. The path to that result is reached through a but-for test: but for the failure to debias for race, Amir would not have received, on account of the AI instrument's prediction of recidivism, a harsher sentence than Jeff. And, as you may have guessed, we can bring the same framework for objectively ascertaining intent to bear on the courts themselves. Suppose we accept that a court has a duty to prevent discriminatory acts or policies from seeping into its proceedings. To discharge that duty, the court needs to keep up with the times. As far as AI-based sentencing instruments are concerned, the court needs to remain reasonably informed of the emerging consensus about the racial biases of the underlying datasets. Knowing what it knows by dint of being reasonably so informed, if the court fails to counteract those biases, it has intended to let those biases prevail. The court itself, setting aside the AI-based instrument, is now suddenly in the business of letting directly discriminating biases in respect of race prevail.

Punishment

There are many headache-saving virtues to ascertaining intent objectively in trying to safeguard legally protected characteristics. The but-for test for intent can be applied against acts and omissions as well as on aggregate or single features. Coupled with dataset debiasing or the outright exclusion of certain features which we know incline AI-based instruments to discriminatory outputs, the but-for test is potent against a wide variety of harms. But let me now close by remarking on a foundational worry about the use of AI-based risk-assessment instruments which returns us to the issue of group to individual inferences reflected in this chapter's opening vignette.

Sentencing is a prelude to punishment and, whatever one's views about the justificatory grounds for punishment per se, no one thinks that we should be sentencing people more harshly on the basis of features of their lives for which they are not blameworthy.

These features can help mitigate but not aggravate sentencing.[19] Now notice that, for all their technical complexity, AI-based predictive instruments are conceptually simple: they take data from the past about a group to make guesses about the future for a particular individual who, in most circumstances, had no causal role in the constitution of that historical data. Thinking about it this way, arguably the only remaining factors which can justifiably feature in the algorithms of predictive sentencing instruments are those for which the individual in question could be blameworthy. That is a radical curtailment of eligible variables, for it essentially excludes, in most circumstances, problems associated with group to individual inferences. That, in my view, wouldn't be such a bad thing—but why the caveat *in most circumstances*?

For one thing, it may well be that an individual did actually play a causal and blameworthy role in constituting the historical data underlying the AI-based instrument's prediction. Take, for example, the vexed issue of prior convictions, which, depending on the individual, may or may not be a product of uneven law enforcement and judicial bias. Debiasing the underlying dataset in these circumstances will be inapt unless we know whether there is, in fact, a problem. Are the individual's prior convictions a result of uneven law enforcement and judicial bias, decision-making as an autonomous moral agent, plain bad luck, or some mix of all of the above? To do justice to the question entails doing justice to the dignity of the person about whom it is posed. That is, to treat the individual as an individual and not a mere datapoint. Of course, there are many ways to do this. The most apt approach, what Kasper Lippert-Rasmussen calls the *revisionist* requirement,[20] doesn't entail discarding generalisations about people but, instead, requires that we supplement and displace those generalisations with fine-grained information about the people we are judging—especially when the stakes are as high as they are in criminal justice. Whether AI is theoretically capable of that kind of epistemic open-mindedness, however, is an open question.

What is not an open question is that the courts must be able to treat individuals as individuals; indeed, they have a duty of fairness to do so. This duty extends to disallowing the use of datasets which they know, or ought to know, will reinforce existing discriminatory biases.

Notes

1. A. Leung, F. Woolley, R. E. Tremblay, and F. Vitaro, 'Who Gets Caught? Statistical Discrimination in Law Enforcement', *Journal of Socio-Economics* 34 (2005): 289.
2. D. Knox, W. Lowe, and J. Mummolo, 'Administrative Records Mask Racially Biased Policing', *American Political Science Review* 114 (2020): 619–37.
3. M. Dhami and I. Belton, 'Using Court Records for Sentencing Research: Pitfalls and Possibilities', in J. Roberts (ed.), *Exploring Sentencing Practice in England and Wales* (Springer, 2015), 19.
4. US Sentencing Commission, *Demographic Differences in Sentencing: An Update to the 2012* Booker *Report* (2017).
5. S. Wortly and M. Jung, 'Racial Disparity in Arrests and Charges', Ontario Human Rights Commission, July 2020.
6. S. Mayson, 'Bias In, Bias Out', *Yale Law Journal* 128 (2019): 2218.
7. K. Hannah-Moffat and K. S. Montfard, 'Unpacking Sentencing Algorithms: Risk, Racial Accountability and Data Harms', in J. W. de Keijser, J. V. Roberts, and J. Ryberg (eds), *Predictive Sentencing* (Hart, 2019).
8. E.g., UK Equality Act 2010.
9. Canadian Human Rights Act (R.S.C., 1985, c. H-6) s. 3.
10. UK Equality Act 2010, s. 10.
11. Giving rise to the 'perfect proxy' problem: B. Davies and T. Douglas, 'Learning to Discriminate: The Perfect Proxy Problem in Artificial Sentencing', in J. Ryberg and J. V. Roberts (eds), *Sentencing and Artificial Intelligence* (Oxford University Press, 2022).
12. K. Lippert-Rasmussen, 'Algorithm-Based Sentencing and Discrimination', in Ryberg and Roberts, *Sentencing and Artificial Intelligence*.
13. T. Khaitan, *A Theory of Discrimination Law* (Oxford University Press, 2015), 75.
14. E.g., *State v. Loomis*, 881 N.W.2d 749 (Wis. 2016) at 766–7 on the non-discriminatory objective of accurately predicting recidivism.
15. 'Beyond Intent: Establishing Discriminatory Purpose in Algorithmic Risk Assessment', *Harvard Law Review* 134 (2021): 1760, 1760–81, <https://harvardlawreview.org/2021/03/beyond-intent-establishing-discriminatory-purpose-in-algorithmic-risk-assessment>.
16. *James v. Eastleigh Borough Council* [1990] 2 All ER 607, Lord Goff at 774e. For a critique of the view that prevailed in *James*, see J. Gardner, 'On the Ground of Her Sex(uality)', *Oxford Journal of Legal Studies* 18 (1998): 167, 186. On

those 'elusive terms', see also B. Hass, 'The Opaqueness of Rules', Oxford Journal of Legal Studies 41 (2021): 407, doi:10.1093/ojls/gqaa054.

17. Khaitan, *A Theory of Discrimination Law*, 162.
18. *Bostock v. Clayton County* 140 S. Ct. 1731 (2020) at 1739, a case which ultimately retained intent. For complications, see S. Atrey, *Intersectional Discrimination* (Oxford University Press, 2019).
19. Sentencing Guidelines Council, *Overarching Principles: Seriousness* (Sentencing Council, 2004).
20. K. Lippert-Rasmussen, '"We are all Different": Statistical Discrimination and the Right to be Treated as an Individual', *Journal of Ethics* 15 (2011): 47.

Further Reading

J. Ryberg and J. V. Roberts (eds), *Sentencing and Artificial Intelligence* (Oxford University Press, 2022).

PART II
POLITICS AND GOVERNANCE

5
The Perfect Politician

Theodore M. Lechterman

In what ways, if any, should we seek to automate political decision-making?

In 2021, researchers at IE University conducted a global survey of the attitudes of citizens toward integrating AI into government.[1] They found that a majority of Europeans (51%) supported replacing at least some politicians with artificial intelligence. Citizens of China showed even more enthusiasm for AI-driven governance, with support for this proposal topping 75% for Chinese respondents. Americans and Brits expressed slightly more scepticism, with only 40% and 31% respectively voicing approval for this idea. Despite intriguing variation across countries, the results suggest a very strong attraction overall to upgrading politics with AI. They also invite deeper reflection. Is artificial political leadership something we have good reasons to want?

Ideas for integrating AI into politics are now emerging and advancing at accelerating pace. Here, I aim to highlight a few different varieties and show how they reflect different assumptions about the value of democracy. We cannot make informed decisions about which, if any, proposals to pursue without further reflection on what makes democracy valuable and how current conditions fail to fully realize it. Recent advances in political philosophy provide some guidance but leave important questions open. If AI advances to a state where it can secure superior political outcomes, leading perspectives in political philosophy suggest that democracy may become obsolete. If we find this suggestion troubling, we need to put the case for democracy on stronger foundations.

Automating Representation?

Presently, several options for integrating AI into politics vie for our support. One proposal seeks to have algorithms run for elected office and compete with human candidates. In 2018, a chatbot named Alisa challenged Vladimir Putin for the Russian presidency,[2] and a bot named SAM was developed with similar intentions in New Zealand.[3] A more radical idea, proposed by physicist César Hidalgo, seeks to provide each citizen with a personalized bot, trained on a citizen's own inputs, that negotiates with other citizens' bots to design and approve legislation.[4] Hidalgo envisions using these bots to reduce or even replace the role of human politicians.

These proposals are species of what we might call augmented electoral democracy. Augmentations of electoral democracy seek to alter the way that citizens are represented in the political process. And they reflect explicit or implicit critiques of conventional representation, where professional human politicians compete for public approval. What is the problem to which augmented representation provides the solution?

Efforts to enable algorithms to compete with human candidates for elected positions focus on the comparative limitations of human politicians, who may be corrupted in various ways and suffer from ordinary human constraints and vices, such as fatigue, ignorance, self-deception, and dishonesty. Automated politicians do not get tired; they are hard to bribe; and they have constant access to large libraries of information, which they can process rapidly and reliably.

Extant demonstrations of this idea have relied on rudimentary chatbots, which allow voters to interrogate a model about positions it has reached on various issues. Since no such candidate has yet succeeded in its bid for election, it remains unclear how well these proposals might perform on the job. But there are strong reasons to be sceptical of electing chatbots to represent constituencies alongside human politicians. In particular, the proposals might be faulted for underestimating the tasks required of elected officials.

The duties of elected office are many and varied. They generally require numerous complex tasks and negotiations with different humans. They are not limited to answering constituent questions, drafting proposed bills, and casting votes on legislation—things that a chatbot might be able to do (though of course with room for debate about how well it could do them). Rather, they involve negotiating with other elected officials to reach legislative compromises, managing staff who perform various auxiliary functions, running re-election campaigns, serving on committees, attending ceremonies, and meeting with different stakeholders to assess all sorts of problems and opportunities. While it is conceivable that an advanced and multifaceted AI agent could fulfil these tasks, and even fulfil them better than most humans, it is inconceivable that a mere chatbot could perform them without human supervision.

Rather than adapt AI to fill a role designed for humans, Hidalgo's proposal—to replace all politicians with personalized bots—seeks to reimagine representation in light of the capabilities that AI offers. Its critique of the status quo focuses not on the natural limitations and moral failings of humans but on the idea of representation itself. Many believe that electoral democracy is simply a pragmatic concession to the problem of scale. True democracy, on this reasoning, is direct democracy, where citizens decide directly on all policies. Until recently, direct democracy has been impractical in large, complex societies, where citizens cannot easily convene and vote on every matter. But technology now makes this possible. By answering surveys and sharing data with personal bots, citizens could train their bots to robustly represent their political preferences. These bots could then negotiate with each other to reach agreements on legislation. In turn, legislation would represent more citizens, as it would be the product of everyone's inputs, rather than the product of officials elected by the majority. It would also represent a much fuller picture of citizens' beliefs and desires, as this data would be fed directly into legislative negotiations, rather than reduced to a simple vote of yes or no for a candidate.

While Hidalgo's proposal offers several advantages, it also reflects some controversial assumptions. Are politicians merely a mechanism for processing our political preferences? Professional politicians may have, or be capable of marshalling, more expertise on political questions than ordinary citizens. Since Plato, many philosophers have regarded politics as an intricate art that requires detailed understanding of legislative procedures, an ability to recruit expertise in virtually any domain, shrewd capacities for strategy and negotiation, and exceptional moral judgement. Delegating the art of policymaking to qualified professionals may result in higher-quality outcomes than if all citizens participated equally in all policymaking decisions. Additionally, electing those with the talents and motivation needed to excel in policymaking enables everyone else to focus their labours where they are most suited, resulting in better individual job satisfaction and a more productive economy. Although electoral democracy contains certain tendencies towards inequality, particularly in countries where campaigns are privately financed, it also contains some important safeguards against it. Direct democracy, whether automated or not, might favour citizens with more resources for effective citizenship, such as education, leisure time, or inherent motivation. Because representatives must compete for the approval of their constituents, they are in principle accountable to all of them equally.

The point here is not to argue that electoral democracy is superior to direct democracy. Rather, it is to confront the normative assumptions that lie beneath proposals for upgrading democratic practices. Elected chatbots might have encyclopaedic knowledge of facts and lack many vices of human politicians. But, at least for the foreseeable future, they will also lack the more advanced skills required of political leaders, such as capacities for persuasion and judgement. Meanwhile, a direct democracy in which all citizens enjoy the assistance of personalized robots might reduce certain inequalities and increase the responsiveness of policies to citizens' preferences. But it contains a high risk of generating low-quality results and widening other inequalities.

Replacing Voting with Data?

Other proposals seek even more radical transformations of conventional politics. For proponents of aggregative democracy, the legitimacy of political power requires that decisions fairly satisfy the preferences of those subject to them. The goal of any political system, according to this school of thought, should be to measure accurately, combine fairly, and duly satisfy the preferences of its subjects. And current polities are seriously defective on this score. Obviously enough, in authoritarian polities, citizens have few channels for expressing their preferences in the first place, and it is the preferences of the few that dominate conditions for the many. In self-described democratic polities, however, the expression of preferences is mainly limited to voting, an activity that occurs infrequently, provides limited information, and is subject to numerous cognitive biases. In many societies, citizens also face barriers to voting, and the integrity of elections suffers regular threats. Powerful groups often find ways to influence which people or issues are put to a vote or to sway how decisions are implemented. Thus, replacing or supplementing conventional methods of translating citizen preferences into policies with AI may offer numerous benefits.

Drawing on science fiction and futurist studies, Jamie Susskind invites us to imagine an alternative to electoral democracy called data democracy.[5] It is based on the idea that data generated throughout our daily lives—our travel patterns, online purchases, search and social media activity, metabolic rates, and so on—reveal a much more comprehensive, finely grained, and updated picture of citizens' wants than votes do. Thus, we might supplement or replace voting with a centralized and dynamic system for measuring and analysing inputs from citizens. The system would collect the vast streams of citizens' digital data and aggregate them to create a detailed profile of public opinion.

Susskind imagines using this data to inform the decisions that politicians make about legislation. But if the purpose of politicians in such a system is simply to process citizen preferences, this processing function might be done more reliably by artificial intelligence. A supercomputer could match data about citizen preferences with data about empirical conditions, enabling it to propose or implement policies for satisfying the general will.

However, augmenting politics in the spirit of aggregative democracy also faces serious criticisms that have long trailed this general conception of democracy. One challenge concerns the difficulty of measuring our preferences. The preferences we reveal through our economic decisions or online activity may offer a more comprehensive picture of our inclinations than votes do. But these apparent inclinations may only weakly correlate with what we truly want. The bingeing, doom-scrolling, rabbit-holing, trolling, flaunting, impulse-buying, and leering revealed in our online behaviour might reflect our vices or indiscretions more than what we value and hope to become. If the data processor could somehow overcome these measurement problems to obtain a more reliable indication of our preferences, it would face a further problem: many preferences we have are simply irrational or unreasonable. Even upon due reflection, many people may prefer, at least in certain ways and at certain times, to dominate or abuse others or sabotage their own wellbeing. Should data democracy give equal treatment to preferences founded on ignorance, confusion, or hatred?

An even more fundamental objection to data democracy would target its neglect of deliberation, a neglect that also figures in Hidalgo's proposal. For many democratic theorists, the ability to engage in public debate with fellow citizens is a vital component of the democratic process. It is only through the public exchange of reasons that we can fully develop and validate our preferences. It is deliberation that helps to separate impulses, prejudices, and mistakes from higher-order desires and beliefs. This perspective suggests that insofar as data democracy would involve shrinking or eliminating the role for deliberation in politics, it represents a fatally flawed conception of democracy.

An Artificial Sovereign?

There is an even more radical possibility for seeking to improve or perfect our politics with AI, a possibility that we might call 'algocracy'. Algocracy is similar to the extreme version of data democracy in that it involves a powerful autonomous system governing on the basis of large data flows and convoluted calculations. However, the algocrat, as I imagine it, would not be bound by popular preferences. Rather, it would ultimately be guided by its own judgements about justice and the common good.

How would the algocrat form these judgements? Fundamentally, it would do so based on comprehensive reading and analysis of all extant digitized content. It would read Confucius and Judith Butler, Darwin and Einstein; it would listen to Shostakovich and Shakira, compare Basquiat to the Dutch Masters, and contrast the *philosophes* with social media influencers; it would analyse every patent, medical database, architectural blueprint, computer program, genome, and restaurant menu.

Naturally, the world's digital repositories contain many biases, as the production and preservation of human knowledge throughout history have tended to favour certain groups and perspectives over others. Additionally, the quantity of recent digital content (much of which is low quality) vastly outnumbers the quantity of works preserved from the rest of human history. But if the algocrat were truly smart, it would be able to reason through these problems. Since it would have read every available publication in feminist and postcolonial criticism, the sociology of inequality, and the mechanics of algorithmic bias, it would be able to consider these critiques when analysing materials. Similarly, it would not weight the remarks of current celebrities more heavily than those of canonical thinkers simply because the former have more likes and shares.

I imagine that the algocrat would also have access to data about citizen preferences, whether reflected directly through votes or indirectly through data. Unlike in data democracy, however, the algocrat would not be limited to aggregating and implementing citizen preferences, whatever they happen to be. Rather, it would use its reasoning powers to judge when preferences are relevant to a decision and when they are not, which preferences are worth considering, and how to fairly resolve conflicts between preferences.

Part of what makes algocracy an interesting thought experiment is that it challenges prominent views in political philosophy about democracy's justification. Contemporary philosophers tend to agree that a well-functioning democracy is the most legitimate way of making coercively binding decisions. Citizens governed by democratic processes have stronger reasons to endorse and comply with laws than citizens governed by alternative processes. But philosophers also disagree about the basis for this judgement, about what exactly makes democracy better than other political systems.

For instrumentalists, democracy is justified because it tends to produce superior results. More reliably than alternative systems, democratic polities tend to respect human rights, resolve conflicts peacefully, achieve prosperity, moderate inequality, and realize the preferences of their subjects. Algocracy poses a significant challenge for instrumentalist justifications of democracy, as it seems plausible that such a system could better achieve many or all the outcomes often associated with democratic systems.

Someone committed to the instrumentalist justification of democracy may resist this conclusion by reminding us that current forms of artificial intelligence face numerous limitations: their epistemic reliability is corrupted by biases, and their vulnerability to hacking and human manipulation makes them untrustworthy governors. Some may go further, claiming that advanced forms of AI will develop a survival instinct that leads them to manipulate humans rather than serve human values or impartial moral principles. But a consistent instrumentalist must admit that, if—or when—these challenges can be overcome, delegating political decision-making to an algocrat would become more than a tempting option: it would become a moral obligation.

Is democracy nothing more than an instrument for producing good results? One may acknowledge that AI could advance to the point where it would produce reliably better policies than humans can hope to achieve by themselves. But for non-instrumentalists, democracy is valuable apart from its consequences. An influential recent version of this position holds that democracy is valuable because it is part of an ideal of social equality. Alternative systems—such as monarchies, aristocracies, and oligarchies—involve granting people different amounts of political power. Granting some people more power than others sends the message that some people are inherently wiser or worthier of consideration than others. By rejecting inequalities in political power, democratic political processes instead affirm a commitment to a society marked by the absence of social hierarchy and subordination. Democracy is a necessary consequence of recognizing one another as moral equals.

However, it is not obvious that the non-instrumentalist position has the resources to fend off the challenge from algocracy. There would be no doubt about the algocrat's intellectual superiority. The algocrat is not a human person or someone with whom we have ongoing social relations. Unlike a dictator or plutocrat, the algocrat would not live among us as a member of human society, demanding or regularly receiving social privileges as perks of power. Thus, even the social egalitarian argument for democracy may be unable to resist the temptations of algocracy. If we can engineer a system of rule that gets dramatically better outcomes and does so without creating arbitrary social inequalities, the case for democracy becomes difficult to sustain.

One response to these observations is to conclude that the case for democracy is weaker than many have thought. If or when the technology for superseding democracy's limitations becomes available, we ought to implement it with all deliberate speed. Democracy, according to this perspective, may soon become technologically obsolete. Another response to these observations is to conclude that democracy is not obsolete but that the arguments in its support need re-examination.

Some philosophers have recently contended that the notions of autonomy and self-determination help to explain what is distinctively valuable about democracy. Even if other systems could more reliably or more robustly advance our preferences or track the demands of justice, and even if these systems could avoid creating or worsening inegalitarian social relationships, these systems would leave something missing. Since the resulting policies would not issue from our own agency, some have argued, we would reasonably find ourselves alienated from the world that they create. The voice that democracy allows us is valuable in part because it allows us to see our social world as of our own making. We would not be able to fully see ourselves in a world ruled by AI, even if it reliably ruled in accordance with our interests.

Precisely how much weight this interest in non-alienation carries is unclear. Many people may be willing to trade it away in exchange for dramatic reductions in injustice that algocracy might offer. It remains possible—and perhaps likely—that democracy's value is a complex combination of factors that cannot be reduced to a simple explanation. As the prospects for integrating AI into political decision-making become increasingly feasible, reflection on the foundations of the democratic ideal take on particular urgency.

Upgrading Political Philosophy for the Age of AI

Might AI, or further advances in AI, offer us the perfect politician? To be sure, there are attractive elements in each of the proposals I have surveyed. They suggest that AI can bring about a richer and more efficient harnessing of information than conventional democratic processes. AI may help to produce outcomes that more fairly represent all citizens, more richly reflect citizen preferences, and are more rational and informed. But each proposal also reflects controversial assumptions about the value of democracy—and its limits. Augmented electoral democracy either underestimates the complexity of a politician's job or else assumes that representative institutions are simply an antiquated solution to the problem of scale. Data democracy assumes that preferences are exogenous to political participation and can be revealed indirectly through behaviour. Algocracy assumes that human agency is a dispensable feature of politics.

I contend that we cannot reach intelligent judgements about how to augment politics without deeper reflection on what makes democracy valuable and how that value can be best operationalized. Recent work in political philosophy provides some essential guideposts, but key questions remain unanswered. The possibility that human agency might someday be separable from politics is something that political philosophers are only now starting to take seriously. To better guide public conversations about the advantages and limitations of integrating AI into politics, political philosophy needs an upgrade.

Acknowledgements

Portions of this chapter expand on ideas that I first explored in 'Will AI Make Democracy Obsolete?', *Public Ethics Blog*, Stockholm Centre for Ethics of War and Peace, 4 August 2021, <https://www.publicethics.org/post/will-ai-make-democracy-obsolete>. I thank the Stockholm Centre for granting copyright permission.

Notes

1. Sam Shead, 'More than Half of Europeans Want to Replace Lawmakers with AI, Study Says', *CNBC*, 27 May 2021, <https://www.cnbc.com/2021/05/27/europeans-want-to-replace-lawmakers-with-ai.html>.
2. Tom O'Connor, 'Will the Next Russian President Be a Robot? Putin's New Challenger Is a Machine That Knows "Everything"', *Newsweek*, 7 December 2017, <https://www.newsweek.com/russia-putin-could-face-controversial-robot-next-year-president-election-741509>.
3. 'Meet the World's First Virtual Politician', *Newsroom*, 15 December 2017, <https://www.newsroom.co.nz/2017/12/14/68648/meet-the-worlds-first-virtual-politician>.
4. César Hidalgo, 'A Bold Idea to Replace Politicians', TED 2018, 5 March 2019, <https://www.ted.com/talks/cesar_hidalgo_a_bold_idea_to_replace_politicians?language=en>.
5. Jamie Susskind, *Future Politics* (Oxford University Press, 2018), 246–50.

Further Reading

J. Susskind, *Future Politics* (Oxford University Press, 2018).

6

Collective Intelligence over Artificial Intelligence

Saffron Huang and Divya Siddarth

Technological progress is a powerful engine of possibility. Penicillin saved hundreds of millions of lives, electricity powers our world, and the internet opened up global knowledge and communication.

But we are in an age of fear and ambivalence about contemporary technological development, particularly about AI. And reasonably so, given its mixed blessings. Better AI algorithms might enable unprecedented access to information or speed up medical research, but they could just as well lead to comprehensive surveillance, or to greater inequality if, for example, a small handful of private corporations create AI that displaces workers worldwide.

AI moves fast; new products and papers are released constantly. Generative AI products (such as GPT-4 and Midjourney) create fluent text, code, images, video, and more. They are shaking up industries left and right, from education to programming.[1] This is driven by profit-maximising innovation dynamics in a few firms: the publications and breakthroughs mostly come from the United States, and often emerge from work at corporate labs like Google.[2] As the generative AI race heats up, these labs are keen to commercialise technology as fast as possible. Developments in AI are not governed by any coherent regulation—unlike, say, new aeroplanes, or new drugs—and there is almost nothing in the way of global coordination such as what we have for nuclear weapons or climate pollution. As it stands, we are all beholden to the incentives of technologists and their funders, and vulnerable to their missteps.

What's more, our current tech industry is not structured to generate broad public benefit from AI. The incentives of venture capital funding models and startup structures push fledgling companies towards high-scalability, high-margin products (such as the common, safe, but not very inspiring work of selling Software-as-a-Service to other companies), rather than more ambitious applications that improve society. Economists are also well aware that markets for innovation have inefficiencies; innovation is a strange good, with highly uncertain outputs, significant spill-over effects into society, and a weird production function.[3] Economists Daron Acemoglu and Pascual Restrepo have noted that these market failures have been particularly salient in AI research and development due to a mix of factors, including the propensity of people to follow the most popular paradigm even if it's less productive than an alternative, and historically low levels of government involvement in encouraging socially beneficial research.[4] Furthermore, in many economically advanced countries, labour is taxed at a much higher rate than capital; this incentivises labour cost-cutting and hence investments in AI applications that try to replace labour. Finally, the most widespread deployment of actual AI thus far is in recommender systems on platforms such as TikTok and Netflix, which at best have had a limited gain for humanity. Ultimately, current economic and social circumstances don't necessarily point AI development in the direction of widespread social benefit.

Imagine if we were all excited by the prospect of AI, because we trust that we live in societies where technology that meets the needs of everyday people is also profitable. In such a world, governments and investors would strive to incentivise and fund AI development that alleviates unpleasant work and augments human capabilities (e.g. investing in augmented reality to help workers, rather than industrial robots to displace them entirely). Additionally, the wealth generated from these advancements would be directed towards broad collective welfare, funding high-quality public education and upgraded transportation networks.

Our social and technological systems are not set up to ensure that transformative technologies will not just *affect*, but *benefit* all of humanity. The core issue of AI governance is figuring out how to systematically restructure how we develop, fund, govern, and distribute the benefits of AI from the get-go. As technology begins to accelerate faster than our bureaucratic democratic governance structures can handle, affecting more people more quickly and more deeply, we need to get better at making decisions in the collective interest. This requires improved institutional design, more effective democratic processes, and smart safety brakes. We need greater *collective* intelligence capabilities to deal with our advancing *artificial* intelligence capabilities.

The Transformative Technology Trilemma

Current approaches to making decisions about AI fall prey to what we may call the *transformative technology trilemma*. They assume the need to accept significant trade-offs between technological progress (advancing technological capabilities), participation (enabling public input and self-determination), and safety (avoiding significant risks). These assumptions can lead to three types of failure— let's call them 'capitalist acceleration', 'authoritarian technocracy', and 'shared stagnation'.

Capitalist **Acceleration**—Sacrificing safety for (technological) progress while maintaining basic participation.

Some people prioritise technological progress over safety, believing in free-market, profit-driven development. Peter Thiel, one of Silicon Valley's most influential entrepreneurs and venture capitalists, personifies this camp. He gave a talk in 2012 at a 'Students for Liberty' conference stating that progress in a particular industry is inversely related to how much government regulation there is.[5] He gave this example: in every decade from the 1500s until the 1980s, ships, planes, and cars made transport faster, and this was only curtailed by big, bad US government regulations. In making this point, he also dismissed the need for society to prioritise protection from climate change and railed against clean tech startups and alternative energy government initiatives.

This is a market-based approach. When it comes to AI, the promise is that technological gains will improve people's quality of life by empowering corporations to provide cheap, abundant goods—the best outcome that capitalism has to offer. We will see AI-enabled everything, from pharmaceutical discovery to international freight, from personal assistants to individualised education. However, this view ignores the significant body of economics literature that shows that markets are inefficient when it comes to innovation; rather than government having to get out of the way, government support is actually needed to create the right kinds and amounts of innovation.[6]

People in the 'capitalist acceleration' camp are usually more interested in the pace of technological or scientific breakthroughs than their sustainable dissemination, or who benefits or is harmed by them. They also seem to believe that the existence of consumer choice is enough to satisfy the criteria of participation, that 'dollar voting' by selecting among a narrow list of products can serve the same function as democratic processes. The type of technological progress that is prioritised here can be said to conflate what is economically viable with what is good or necessary, while forgetting what economists know about innovation. Under free-market development there are often negative externalities. For example, the invention of aerosols with chlorofluorocarbons (CFCs) led to the rapid destruction of the Earth's ozone layer; a relatively small number of people created and sold the products, but they socialised the risk of cancer, eye damage, and other problems worldwide. (CFCs were eventually globally banned in 1987.) Clearly, market-driven technological progress isn't sufficient for a flourishing society.

If AI continues to be developed by unconstrained profit-seeking corporations, profits may be prioritised at the expense of other goals. Again, recommender systems that operate as part of Facebook, TikTok, or X (formerly Twitter) are the most widely deployed AI systems today, used to minimise the role of active human judgement in what we read, and to maximise ad revenue and addiction. There is no guarantee that future AI systems will be commercialised for better ends. Continuing to automate cognitive tasks with only money-making objectives in mind could lead to potentially catastrophic impacts.

Authoritarian Technocracy—Sacrificing participation for safety while maintaining basic technological progress.

Some people, in contrast, are particularly worried about risks from advanced AI capabilities. They're kept up at night by fears of misuse of AI by terrorist groups or regular citizens, or the possibility of future catastrophic accidents from over-trusting AI mechanisms with our transport network, weapons systems, or supply chains. Some end up thinking that the only way to ensure safety from catastrophic risks is by restricting the development of AI to a few trusted entities (individuals, companies, or nation-states). They usually assume, too, that collective participation in these kinds of decisions is too dangerous, difficult to coordinate, slow, or likely to lead to lower-quality decisions.

One eyebrow-raising example is the 'vulnerable world hypothesis' from controversial philosopher Nick Bostrom.[7] He postulates that there exists a level of technological development, beyond which technology is so powerful and risky that it's very likely that it will destroy civilisation. Consequently, we need 'a system of ubiquitous real-time worldwide surveillance' to guard against threats from highly dangerous, but as yet unknown technology before it's invented or deployed. This includes creating the capacity for 'extremely effective preventative policing' and 'strong global governance', including fitting everyone with a mandatory 'freedom tag' (we assume that this looks something like a parole ankle tag). This tag would collect all sorts of personal data for review by 'freedom officers', which sounds like a Newspeak term straight out of *Nineteen Eighty-Four*. To top it off, Bostrom seems to believe in the benefits of 'pre-emptive incarceration', dismissing the downsides of locking innocent people up and the potential for abuse of power.

While few go to such extremes, a lot of people believe in making top-down decisions intended for the good of the people. In many countries during COVID, there was controversy around mandatory lockdowns and vaccines, sparking debates around the trade-off between safety and participation (or liberty). As AI becomes more advanced, the demands of 'safety' may increasingly come to outweigh those of participation.

This is a technocracy-based approach. It assumes the best outcomes come from expert decision-making and monitoring, operating within stable institutional structures. It might make the case for allowing only one or two AI research labs to build advanced technology. Perhaps those labs should be nationalised so that the government can tightly control development. Or perhaps a lock should be put on computer chips so that only certain verified entities can run neural networks.

If AI development takes this path, maybe we'll better understand and be able to mitigate certain risks. But this approach itself risks the basic injustices of autocracy; not to mention, tightly controlled, top-down governance structures can be brittle and inhospitable to valuable bottom-up information flows, which can, as in the case of Soviet-era central planning, ultimately bring down the entire political and economic system.

Shared Stagnation—Sacrificing progress for participation while maintaining basic safety.

This path prioritises participation over progress, while maintaining a level of safety. People in this camp believe that there are consequences to how we pursue progress (such as the acceleration of climate change, ecological destruction, greater inequality, or the erosion of good jobs). In light of this, many desire a return to local-scale decision-making, such as more forms of direct democracy, local production of goods and services, and the explicit or implicit goal of forestalling technological advances. There is a lot of variety in this camp; for example, people in the degrowth movement have called for shrinking rather than growing economies to achieve social equality and sustainability, through initiatives like cutting working hours to reduce income inequality, promoting local production, and moving beyond GDP as a measure of progress. In AI, some have called for a slowdown in the speed of AI advancement, while others have looked to endorse participatory processes that involve communities in AI development.[8]

This is a democracy-based approach. It emphasises equality and shared power, and starts from the assumption that equal participation is necessary for just outcomes. There is much to like here, but significant problems remain. Many of the methods proposed seem inadequate to stop the big companies and projects from ploughing

ahead with AI, they ignore the many people and countries who welcome such innovation, or they overlook the lack of interest of many
in participating in time-consuming decision-making processes.
Furthermore, there are large-scale projects that could be greatly
helped by AI technologies in areas from clean energy, to public
health, to public transport; impeding progress in AI might therefore come at a significant cost.

Undeniably, the methods we currently have for enabling broad
participatory input are too inefficient and inadequate to be implemented widely. Standard electoral democracies are not equipped to
capture high-fidelity information about the will of the people; they
generally rely on the lossy signal of one vote from each person every
few years (and even then, many can't be bothered to vote), alongside the influence of pressure groups and media. Even if politicians
did know what people want, many blockers sit in the way of turning
those wants into relevant outcomes (e.g. special interest groups that
create regulatory capture). More local-level, information-rich processes such as town halls tend to be too strenuous for regular people
to engage in. Democracy is a necessary foundation for any just
model of AI governance, but the versions of democracy we are currently working with are not nearly up to the task.

Who Decides Who Decides?

Markets, technocracy, or democracy? Each of these approaches to
AI, pursued exclusively, gives up too much; we should be aiming for
a mixture of all three. We want to develop social systems that
encompass progress, safety, *and* participation.

In reality, technological outcomes tend to be unevenly spread
among the population. Some inevitably suffer collateral damage,
while others will experience improvements in their lives. Some will
be safe, and others will be put at risk. Some will have their voices
heard, others will not. And it's often difficult to know in advance
who will benefit. Take the Scottish Highland Clearances in the 18th
and 19th centuries during which many tenants were evicted. This
followed agricultural modernisation and the development of the

textile industry, and led to landowners converting land away from traditional farming to sheep farming. The landowners benefited, but the tenants did not—many Highlanders suffered displacement, poverty, and emigration. Similarly, modern-day ride-sharing apps have benefited many ride-hailers and drivers looking for flexible work, but negatively impacted the taxi industry and created concerns around the gig economy.

Given the uncertainties of technological processes, we need ways to decide trade-offs between progress, safety, and participation, and to determine who is affected and in what ways. We should make trade-offs with care and state-of-the-art information-gathering.

So, how can we make governance decisions about AI in our collective interest? This requires two things. First, ways of determining what is in the collective interest, including what kinds of safety might we be happy to trade off for specific kinds of progress, and what modes of decision-making we prefer in certain contexts. Second, ways of implementing the decisions made in the collective interest. These are not easy issues; in fact, they're core components of both economics and political philosophy. AI governance problems are a microcosm of governance problems as a whole, mirroring the question of what mixture of market-driven, technocratic, or democratic governance is best for society. Getting better at governance is central to our future, so that we can make increasingly consequential decisions about increasingly consequential technology.

The Case for Better Collective Intelligence

We believe that a new approach to governance is possible through not just focusing on creating *artificial* intelligence, but embracing and developing our *collective* intelligence (CI). If we can integrate and coordinate people's intelligences better, we can give people real agency in governance and also make better decisions with higher-quality information. The tools of collective intelligence consist of technologies, processes, and institutions that can incorporate collective participation at scale to make collective decisions.

There are three components to CI: collective cognition, collective cooperation, and collective coordination. Collective cognition involves aggregating beliefs, goals, values, and preferences from the collective. Collective cooperation is synchronising actors and activities towards a common goal. Lastly, collective coordination is enabling maximally desirable shared outcomes in a world where, inevitably, preferences and values often don't align.

An example of a collective intelligence initiative that served all three functions is the Pol.is collective deliberation software that has helped find public consensus at scale on tech regulation in Taiwan and elsewhere. The issue at hand in Taiwan in 2015 was whether Uber should be allowed into the country. The public liked the convenience, but taxi drivers were losing out and there were operational controversies, such as whether Uber should be obeying Taiwan's laws for transport companies or for tech companies. Benefits and losses from this new app service were distributed unevenly. So the government asked g0v, a group of 'civic hackers', to bring together citizens and other stakeholders to discuss the issue online over a period of weeks on a new platform, which crucially used the Pol.is tool to conduct discussions.

The Pol.is open-source software tool allows people to input their views and decide whether they agree or disagree with others' statements, ultimately creating a useful map of consensus. What camps do people coalesce into, and do they overlap? Which statements do specific camps generally agree or disagree on, and which are agreed upon across camps?

In the Taiwan case, two large groups emerged, one that was pro-Uber and a larger one that was anti-Uber. To incentivise the creation of viable solutions, the organisers decided that for any given recommendation to be considered, at least 80% of the participants needed to agree with it. Over time, and with more discussion, six recommendations emerged that received over 80% approval; Pol.is's map of consensus shifted accordingly. These recommendations included that Uber needed to put forward a good-faith plan to pay taxes in Taiwan, and that Uber drivers needed to display their registration certificate, license, and drivers' information visibly in the car. More meetings were held with a smaller, but still representative subset of stakeholders, to flesh out the results of the crowdsourced Pol.is discussion. Finally, a number of proposals were ratified in 2016, the process improved the relationship between Uber and traditional taxi drivers, and Uber agreed to a number of measures, with the sole exception of becoming a taxable entity. This process has now been used to pass twenty more pieces of legislation in Taiwan.

Pol.is not only lets participants shape the conversation by inputting views in their own words, it also shows both the organisers and the participants where consensus exists on issues, sometimes unexpectedly. In the words of the founder Colin Megill, the tool 'gamifies consensus', rather than other online social platforms that tend to exacerbate polarisation and disagreement. In the Taiwan case, it allowed collective cognition, as it created an easy way to aggregate and visualise thousands of different views, stated in people's own words; it provided a landscape of opinions, beliefs, goals, and values of the community. As a system, Pol.is makes collective cooperation possible, since people can find others with the same (or different) views in the opinion maps, understand what others in their cluster agree with, and start to propose recommendations for action. Lastly, Pol.is enables collective coordination, since it lets a disparate group of stakeholders come together and, while not agreeing on everything, figure out what others believe and why, and what the most optimal paths forward might look like.

Many tools, mechanisms, and systems have been invented to enable collective intelligence.[9] Small focus groups, large crowdsourcing efforts, and prediction markets enable collective cognition by surfacing different kinds of useful information. Team management tools and new corporate structures help people row in the same direction for collective cooperation; for example, 'focused research organisations' are new types of non-profit entities focused on collaboratively advancing research in a specific scientific field for a time-bounded period.[10] Sort of like a strike team for science. Additionally, radical funding mechanisms such as 'quadratic funding' help to level the philanthropic playing field.[11] Quadratic funding uses a mathematical formula that allocates matching funds, more heavily weighting projects that get lots of small donations than those that get fewer, larger donations. This ensures that everyone's contributions count, regardless of their financial means, and promotes projects with broad support.

When it comes to AI, we need ways for people to participate collectively in negotiating complex value trade-offs with global consequences, and ways to act on collective preferences. When trade-offs are made, they need to be made in light of state-of-the-art information-gathering, not preconceived assumptions. We can incorporate collective perspective-gathering systems at different stages of technological development and in different communities, organisations, and constituencies. This would ensure AI accurately represents public opinion, and that AI decision-making involves more democratic feedback loops. We also need the right incentives and resources in place so that socially beneficial AI projects are funded, the benefits of AI are fairly distributed, and we mitigate or compensate for the risks.

AI itself can be used to support collective intelligence. For example, imagine an AI-augmented citizen's assembly that enables collective input to understand what the current global impacts of generative AI look like. This might convene people from all around the world, supported by AI-enabled translation capabilities, to discuss the current impacts and paths forward—and why they differ across cultures and countries. A large language model AI might ask participants a series of customised questions to accurately elicit their preferences, and to make participation engaging. Another AI model could help decision-makers understand the outcomes of global-scale, mass deliberations by representing complex, large-scale information in digestible ways. After decisions are made, an AI-enabled platform could marshal news relevant to different communities given the transcript of the discussions, to help citizens hold technology companies accountable.

AI algorithms could even help us find proxy measures for 'the common good' that encompasses some mixture of progress, participation, safety, and other goals. They could also design new corporate structures to find new models of organising people towards creating products and services for the common good.

Making our societies more collectively intelligent is necessary for ensuring that AI and other transformative technologies will broadly benefit people (again, imagine if we had faith that this would happen by default!). Not only is this necessary, it is also increasingly possible. As AI accelerates, the time is ripe for greater collective intelligence, and the better futures we could bring to life.

Notes

1. T. Eloundou, S. Manning, P. Mishkin, and D. Rock, 'GPTs are GPTs: An Early Look at the Labor Market Impact Potential of Large Language Models' (arXiv:2303.10130), <http://arxiv.org/abs/2303.10130> (2023); D. Milmo, 'ChatGPT Allowed in International Baccalaureate Essays', <https://www.the-guardian.com/technology/2023/feb/27/chatgpt-allowed-international-baccalaureate-essays-chatbot>; S. Peng, E. Kalliamvakou, P. Cihon, and M. Demirer, 'The Impact of AI on Developer Productivity: Evidence from GitHub Copilot' (arXiv:2302.06590), <http://arxiv.org/abs/2302.06590> (2023).

2. D. Zhang, S. Mishra, E. Brynjolfsson, J. Etchemendy, D. Ganguli, B. Grosz, T. Lyon, J. Manyika, J. C. Niebles, M. Sellitto, Y. Shoham, J. Clark, and R. Perrault, *The AI Index 2021 Annual Report* (Human-Centered AI Institute, Stanford University, 2021).

3. K. Bryan and H. Williams, 'Innovation: Market Failures and Public Policies' (No. w29173; p. w29173). National Bureau of Economic Research, <https://doi.org/10.3386/w29173> (2021); U. Witt, 'Innovations, Externalities and the Problem of Economic Progress', *Public Choice* 89/1–2 (1996): 113–30.

4. D. Acemoglu and P. Restrepo, 'The Wrong Kind of AI? Artificial Intelligence and the Future of Labor Demand'. National Bureau of Economic Research Working Paper (2019).

5. Students For Liberty, 'Peter Thiel at the ISFLC 2012', 26 February 2012, <https://www.youtube.com/watch?v=k3rp4jXTYJU>.

6. K. Arrow, 'Economic Welfare and the Allocation of Resources for Invention', in *The Rate and Direction of Inventive Activity: Economic and Social Factors* (Princeton University Press, 1962), 609–26; K. Bryan and H. Williams, 'Innovation: Market Failures and Public Policies' (No. w29173; p. w29173), National Bureau of Economic Research, <https://doi.org/10.3386/w29173> (2021).

7. N. Bostrom, 'The Vulnerable World Hypothesis', *Global Policy* 10/4 (2019): 455–76, <https://doi.org/10.1111/1758-5899.12718>.

8. A. Birhane, W. Isaac, V. Prabhakaran, M. Díaz, M. C. Elish, I. Gabriel, and S. Mohamed, 'Power to the People? Opportunities and Challenges for Participatory AI', *Equity and Access in Algorithms, Mechanisms, and Optimization* 1–8 (2022), <https://doi.org/10.1145/3551624.3555290>; S. Samuel, 'The Case for Slowing Down AI', *Vox* (13 March 2023), <https://www.vox.com/the-highlight/23621198/artificial-intelligence-chatgpt-openai-existential-risk-china-ai-safety-technology>.
9. G. Mulgan, *Big Mind* (Princeton University Pres, 2017).
10. A. Marblestone, A. Gamick, T. Kalil, C. Martin, M. Cvitkovic, and S. G. Rodriques, 'Unblock Research Bottlenecks with Non-profit Start-ups', *Nature* 601/7892 (2022): 188–90, <https://doi.org/10.1038/d41586-22-00018-5>.
11. E. Posner and G. Weyl, *Radical Markets* (Princeton University Press, 2018).

Further Reading

J. Henrich, *The Secret of Our Success* (Princeton University Press, 2017).

PART III

WORK AND PLAY

7

Work and Meaning

A Challenge for Economics

Daniel Susskind

Introduction

Every day, we hear stories of systems and machines taking on tasks that until recently we thought only human beings could ever do: making medical diagnoses and driving cars, drafting legal documents and designing buildings, composing poems and writing news reports. Few areas of life have been unaffected. And what these technological achievements have in common is that they are all driven, in large part, by advances in AI.

In the field of economics, this progress in AI has challenged the traditionally rigid distinctions that economists use to distinguish between tasks that can and cannot be readily automated: in particular, the distinction between 'routine' tasks that can be automated and 'non-routine' that cannot.[1] And these advances have undermined long-standing assumptions about the sorts of jobs that are protected from automation: white-collar roles, for instance, were thought to be out of reach due to the 'non-routine' activities they involve (such as tasks that require creativity and judgement) but that turns out to have been a mistake.[2]

As a consequence, the economic literature is now characterised by far greater agnosticism with respect to the limits of machine capabilities. There are fewer attempts to demarcate a fixed boundary marking which tasks machines can and cannot do, and fewer strident claims about exactly which particular jobs are most at risk of automation. Instead, there is a growing recognition that these technologies, driven by advances in AI, are becoming increasingly capable over time, albeit in directions that are difficult to predict.[3] A process of 'task encroachment' is now underway whereby machines gradually, but relentlessly, take on more tasks that once fell to human beings.[4]

Unsurprisingly, economists have almost exclusively focused on the economic impact of these new technologies on the labour market: in particular, the implication of this process of task encroachment for employment and earnings. Until recently, a benign view of automation prevailed in the field: the belief that new technologies would broadly benefit workers. But over the past few decades, that position has come under attack. The latest research, with its revised conception of machine capabilities, has contributed to a creeping pessimism in the discipline. And again, it is AI that is largely responsible for this shift in sentiment.[5]

Importantly, though, the impact of AI on the labour market is not limited to this economic dimension. It is often said that paid work is not only a source of income but of 'meaning' as well. The term 'meaning', when used in this way, is handled inconsistently, in both formal and informal commentary. In this chapter, I use the term to capture the way in which work provides some people with a source of purpose, an opportunity for self-fulfilment, and a sense of structure and direction in life. (From now, I also refer to 'paid work' as simply 'work'.) And if that broad claim is right, then the challenge of further technological progress, driven by ongoing advances in AI, is not just that the labour market might be hollowed out, leaving some workers without work or with a different type of work, but that it might hollow out that sense of meaning as well.[6]

Over the centuries, this relationship between work and meaning has attracted the attention of some of the great scholars in philosophy, sociology, social psychology, and elsewhere.[7] But in this chapter, I want to focus on how *economists* in particular have thought about it: in the first half, I'll set out the contradictory and haphazard way that they have engaged with this relationship in the past; and in the second, I'll explain why they must take this relationship far more seriously in the future.

Work and Meaning in Economics

Traditionally, the economic literature has handled the relationship between work and meaning in a very narrow way. In the textbooks, work effort is necessarily a source of disutility, and that effort is only provided by a worker in return for a wage. This approach has a rich history, stretching back to Adam Smith's conception of work as a source of 'toil and trouble'.[8] Of course, in the broad intellectual history of economic thought there have been some deviations from this setup. Among the classical economists, for instance, Alfred Marshall famously claimed that 'man rapidly degenerates unless he has some hard work to do, some difficulties to overcome'—in his view, work was not only a source of income but the way to achieve 'the fullness of life'.[9] And among contemporary researchers, there are 'behavioural' and 'happiness' economists who have treated this relationship between work and meaning more rigorously.[10] But these efforts are exceptions not the norm: 'in economics' stated one survey of the field, 'there has been relatively little discussion about the desire for 'meaning''; 'the idea that work has meaning and is meaningful beyond its contribution to personal consumption', wrote another, 'has been largely absent from mainstream economics'.[11]

Nevertheless, it is unlikely that most economists really believe that, in practice, work is a source of disutility alone, and that such a strictly negative relationship between work and meaning actually holds; as Betsey Stevenson put it, 'Most economists are concerned about how we allocate jobs and underneath that concern lies a belief that work matters independent of the earnings that are generated by the work.'[12] One can see this in many of the public comments made by leading labour economists. David Autor, for instance, has argued that 'Idleness is a terrible thing' whereas 'Work gives people's lives structure and meaning.'[13] And in a similar spirit, the MIT Task Force on the Future of Work argued that 'work provides, in the best case, purpose, community, and esteem to those who engage in it'.[14] Here, Autor and others are making the simple claim that work is not just a disutility-inducing means to a wage, as the traditional models assume, but a means to other valuable ends as well.

Other economists have made yet stronger claims about the relationship between work and meaning—that work is not just a sufficient condition for meaning, but a necessary one. Daron Acemoglu, for example, has made the case that 'it is good jobs, not redistribution, that provide people with purpose and meaning in life', and unless we create 'meaningful employment' for most people they will lack a 'viable social purpose'.[15] This might even hint at new ways to measure and assess the economy: Robert Shiller argues that since 'jobs are more than a source of income' it follows that the 'objectives of our discipline [economics]…shouldn't be the GDP or productivity number so much as the meaning of life number'.[16]

So there is a conflict between the healthy relationship between work and meaning appealed to by economists in public commentary, and the harmful one reflected in formal models. Highlighting this conflict is not necessarily intended as a criticism. Model building is always an exercise in simplification, and certain features of the real world must be left out to make the model easier to handle. The central question is whether this lack of modelling realism is sufficiently consequential for the particular outcomes we are interested in studying. In the past, where the focus of these models was overwhelmingly on the economic dimension of work, this lack of realism was comparatively inconsequential. However, it has become more conspicuous as our collective focus increasingly shifts to the non-economic dimension of work.

Taking the Relationship Seriously

There are three reasons for economists to take the relationship between work and meaning more seriously than at present.

To begin with, it appears to influence important economic outcomes. Of course, it is well established that non-monetary features of a job might affect labour market outcomes. Any undergraduate student of labour economics, for example, will be familiar with the 'fair wage' or 'efficiency wage' hypotheses, where workers concerned with 'fairness' are more willing to exert effort in response to a higher wage (because they perceive a higher wage to be fairer).[17] And as far back as Adam Smith, scholars have argued that the non-economic features of jobs could compensate for variations in wages; in Smith's words, they could 'make up for a small pecuniary gain in some employments, and counter-balance a great one in others'.[18] In economics today, this idea of 'compensating differentials' is standard.[19]

But despite this general recognition of the importance of non-monetary and non-economic aspects of work—a relationship that was present well before AI made it so pressing—there has been little formal research on the link between work and meaning specifically. And this matters because, as noted, this dimension of work can determine outcomes that traditionally interest economists. For example, perceptions of meaningful work—defined in this case as involving activities that 'individuals view as purposeful and worthwhile'—are said to predict 'retirement intentions, absenteeism, and skills training'.[20] Given such results, the fact that many workers do not appear to gain a sense of meaning from their work is significant: in the US, almost 70 per cent of workers are either 'not engaged' in or 'actively disengaged' from their work; in the UK, almost 40 per cent of people think that their work does not make a meaningful contribution to the world.[21]

The second reason that economists ought to be more concerned with the work–meaning relationship is that it is relevant for understanding how technological change, driven on by AI, affects not only the quantity of work, but its *quality* as well. Popular commentary on the future of work has tended to focus on the former—on the number of 'jobs' that have to be done, which 'jobs' are most at risk of automation, whether there are going to be enough 'jobs' for people to do. In that spirit, one influential study is regularly reported as claiming that 47 per cent of US jobs are at risk of automation in the coming decades.[22] But this focus is too narrow: there are many other ways in which the labour market can adjust to technological change, besides a change in the number of jobs.

In contrast to journalists and commentators, academic economists have traditionally been concerned with those other important effects: less focused on how technological change affects the number of jobs and more on the nature of those jobs. Consider the following indicative statements. David Autor: 'Even if automation does not reduce the quantity of jobs, it may greatly affect the qualities of jobs available.' Laura Tyson and John Zysman: 'We are skeptical that AI and ongoing automation will support the creation of enough good jobs.' And Andrew McAfee paraphrasing Robert Gordon: 'we don't have a job quantity problem, we have a job quality problem'.[23] There is a growing focus on how technological progress can be 'directed' in a way that promotes 'good' work, and a strong emphasis on how the present path of automation is not inevitable.[24]

What is meant, though, by the 'quality' of work? What does 'good' work involve? A variety of criteria have been used. The most common among economists is the wage level. Another is job 'security'.[25] Yet defining a job as 'good' according to these narrow aspects alone will lead to a shallow conception of 'good' or 'high-quality' work. A richer conception must engage with the idea that work is also a source of meaning. And in thinking about AI, this task is particularly important, given the impact it has on the quality of work. To start with, as we saw before, new technologies are now taking on 'non-routine' tasks. This matters because these are often thought to be more meaningful activities: creating interesting images, designing beautiful objects, solving problems that require subtle intellectual faculties. In turn, the tasks that are hardest to automate—often involving manual dexterity or interpersonal skills—are found in low-paid, low-skill, insecure service roles, not the sorts of jobs that self-evidently provide meaning. And then, there are the ways in which AI itself makes work less pleasant: for instance, by supporting more intensive worker surveillance, more precise performance target setting, and more insecure on-demand shift scheduling.[26]

The significance of having a broad view of 'good' work is increasingly recognised by philosophers of work: Joshua Cohen, for example, distinguishes between the 'standard goods' of a 'good' job, which include 'compensation' and 'stability', and the more 'ambitious' aspects like 'purpose'.[27] Economists too have internalised the relevance of 'ambitious' attributes, as evidenced by their public comments on the importance of the relationship between work and meaning. Clearly, if they are to make a substantive contribution to debates about the impact of AI on the quality of work, economists cannot simply focus on technical issues like pay and flexibility. But it is also right that claims about the work–meaning relationship are subject to the same critical attention as those about more traditional economic questions. This requires careful engagement with the nature of this relationship.

Another reason that economists ought to take the relationship between work and meaning more seriously concerns the longer term. As noted at the start of this chapter, because of recent developments in AI, a shift has taken place in the economic literature—towards a far more pessimistic view of the impact of technological change on work. And in some scenarios, that progress not only affects the quality of work, but the *quantity* of work that must be done as well. In response, policymakers, economists, and politicians have begun to explore the more radical interventions that might be required to respond to the challenge of there not being enough good work to go around. The two interventions that have attracted the most attention in the UK (and elsewhere) are the 'Universal Basic Income' (UBI) in which everyone receives a cash grant with no strings attached, and the 'Job Guarantee Scheme' (JGS), in which everyone is provided with paid work, funded by the state.[28] At present, though, the broader debate about their merits is clouded by a failure to explicitly engage with the dual purpose of work: as a source of income *and* meaning.

To see this limitation, it is useful to distinguish between the two distinct problems that a society would face in a world where there is not enough work. First there is the economic problem: how to provide people with an income. This is a problem of distribution which, until now, has predominantly been solved through paid work. If the labour market is a less reliable option in the future, though, then an alternative distribution mechanism will have to take its place. But there is also a meaning problem: how to provide people with meaning if work no longer sits at the centre of their lives. Given the heterogeneity in the nature of the relationship between work and meaning—the simple fact that many people don't get a sense of meaning from their work, noted before—this problem will be more acute for some than others.

This decomposition of the challenge of the workless is useful for thinking clearly about the relative appeal of interventions like a UBI or a JGS. With respect to the economic problem, both interventions are similar. Both the UBI and the JGI provide people with an income, independent of their status in the labour market: the former does it on account of membership in the political community (for that reason it is sometimes called a 'citizen's income'); the latter does it in return for a state-provided job (i.e. one that would not otherwise be provided by the labour market). From an economic point of view, the differences between the two interventions are technical rather than substantive. However, these interventions differ more markedly with respect to the meaning problem. This is because they tend to be based on different assumptions about the nature of the relationship between work and meaning.[29]

If one believes that work and meaning are linked, then a JGS is likely to be appealing—not only does it provide an income, but by providing some sort of state-supported work it also restores the sense of meaning that a person might have got through traditional work in the labour market. In contrast, if one is more sceptical about the strength of the relationship between work and meaning, or is more confident that people would be able to find meaning in non-work activities, then a UBI is likely to be preferred—it provides people with an income but allows them to find meaning through those other activities instead. Advocates of these different interventions rarely make this distinction between the different problems that are being solved. But implicitly, this is often what is taking place.

Consider the economist Daron Acemoglu's concern about interventions like UBI as a response to the challenge of automation: 'Building shared prosperity based predominantly on redistribution is a fantasy,'[30] he has written, and 'no society has achieved shared prosperity by just redistributing income from the rich to the less fortunate.'[31] But the weight of his argument is less on technical concerns, that a UBI-like scheme would be large and unwieldy, and more that such a scheme would fail to solve the meaning problem: without work, people would be without 'a viable social purpose'[32] and 'it is unlikely that individuals could find a similar meaning or purpose from pure redistribution, no matter the scale'.[33]

The point here is not to take a position on the merits of UBI, but instead to show how arguments about the bold interventions that might be required often rest on two tacit assumptions: that work is a source of meaning and that, in turn, non-work activities are a poor alternative source of meaning.

These assumptions cannot be taken for granted. The first relationship—between work and meaning—is challenged by the heterogeneity described before (again, some but not all get meaning from their work). The second relationship—between non-work activities and (lack of) meaning—is challenged by the intuitive observation that many people do in fact find meaning outside of the formal labour. market, not only in hobbies and recreational activities, but also in the great variety of work that goes unpaid—care services, volunteering, work in the domestic economy. Engaging with the truth of these assumptions is hard, given the fact that work currently sits at the centre of most people's lives—not least among those who write about this subject—and given the dominance of the work ethic that determines what it means to be a valuable member of society today, it is difficult to imagine how things might be done differently. But if technological progress is carrying us towards a world that is quite different from our own, where there is not enough demand for the work that human beings do to keep everyone in a good job, then we must confront the nature of the work-meaning relationship more seriously.

The line of investigation in this chapter is likely to be unfamiliar to most economists who explore the impact of technological change on the labour market. The failure to investigate this using formal economic methods was understandable in a world where the impact was relatively muted. However, given the encroachment of new technologies on an ever-widening range of tasks, driven by advances in AI, and a recognition that the effect may be far more disruptive than previously imagined, we now need a sustained focus on work's non-economic dimension. Some economists have begun to capture the meaning that work might provide people in their formal models; Anton Korinek, for instance, includes a measure of meaning in the utility functions of workers, and then applies conventional economic analysis to that unconventional economic setup.[34] This is a promising direction of travel. But unless these sorts of investigations are taken up more widely, economists will be forever stuck with a very thin conception of work. If in future economists ignore the issue of meaning, their attempts to understand work by only studying wages will be much like trying to understand a piece of music by only studying the notes on the page without listening to the sound it makes when performed—dry, colourless, and incomplete.

Acknowledgements

This chapter builds on D. Susskind, 'Work and Meaning in the Age of AI', Brookings Institute Working Paper, 19 January 2023—particularly Section 1 and 2. It also draws on D. Susskind, 'Artificial Intelligence, Liberal Neutrality, and Rawls' (forthcoming). Thank you to all participants at the Institute for Ethics in AI seminar at Oxford University, and in particular to David Edmonds for such useful editorial advice. This research was part funded by the Economic and Social Research Council and the Japan Society for the Promotion of Science (Grant number ES/W01159X/1).

Notes

1. See e.g. D. Susskind, 'Re-Thinking the Capabilities of Technology in Economics', *Economics Bulletin* 39/1 (2019): 280–8; D. Susskind, *A World Without Work: Technology, Automation, and How to Respond* (Allen Lane, 2020).
2. D. Susskind and R. Susskind, *The Future of the Professions* (Oxford University Press, 2015).
3. See e.g. in D. Acemoglu and P. Restrepo, 'The Race Between Man and Machine: Implications of Technology for Growth, Factor Shares, and Employment', *American Economic Review* 108/6 (2018): 1488–542.
4. See Susskind, *A World Without Work*; D. Susskind, 'A Model of Task Encroachment in the Labour Market', Oxford University Working Paper (2020).
5. See e.g. Acemoglu and Restrepo, 'The Race Between Man and Machine'; D. Acemoglu and P. Restrepo, 'Modelling Automation', *AEA Papers and Proceedings* 108 (2018): 48–53.

6. See Susskind, *A World Without Work* for the first discussion of this idea. I am grateful to Michael Sandel for the 'hollowing out' analogy.

7. See Susskind, 'Work and Meaning in the Age of AI', for an overview.

8. Cited in Susskind, *A World Without Work*, 216.

9. Cited in Susskind, *A World Without Work*, 216.

10. See e.g. G. Loewenstein, 'Because It Is There: The Challenge of Mountaineering for Utility Theory', *Kyklos* 52/3 (1999): 315–43.

11. L. Cassar and S. Meier, 'Nonmonetary Incentives and the Implications of Work as a Source of Meaning', *Journal of Economic Perspectives* 38/3 (2018): 215–38; and D. Spencer, 'Developing an Understanding of Meaningful Work in Economics: The Case for a Heterodox Economics of Work', *Cambridge Journal of Economics* 39/3 (2015): 675–88.

12. B. Stevenson, 'Artificial Intelligence, Income, Employment, and Meaning', in Ajay Agrawal, Joshua Gans, and Avi Goldfarb (eds), *The Economics of Artificial Intelligence: An Agenda* (University of Chicago Press, 2019).

13. C. Wellisz, 'Late Bloomer', *IMF Finance & Development* 54/4 (2017).

14. D. Autor, D. Mindell, and E. Reynolds, 'The Work of the Future: Shaping Technology and Institutions', MIT Work of the Future Report (November 2019).

15. D. Acemoglu, 'Where Do Good Jobs Come From?', *Project Syndicate* (26 April 2019); D. Acemoglu, *It's Good Jobs, Stupid*, Economics for Inclusive Prosperity Research Brief (2019); D. Acemoglu, 'AI's Future Doesn't Have to be Dystopian', *Boston Review*, 20 May 2021.

16. Robert Shiller, 'Jobs are More than a Source of Income', Uploaded on 14 March 2019. YouTube video, 2:00 min. CEPR & VideoVox Economics (2019); Robert Shiller, 'Narratives About Technology-induced Job Degregation Then and Now', Cowles Foundation Discussion Paper No. 2168 (2019).

17. G. Akerlof, 'Labor Contracts as Partial Gift Exchange', *The Quarterly Journal of Economics* 97/4 (1982): 543–69; G. Akerlof and J. Yellen, 'The Fair-Wage Effort Hypothesis and Unemployment', *The Quarterly Journal of Economics* 105/2 (1990): 255–83.

18. Adam Smith, *An Inquiry into the Nature and Causes of the Wealth of Nations* (Oxford World's Classics, Oxford University Press, 1998).

19. E. Lazaer, 'Compensation and Incentives in the Workplace', *Journal of Economic Perspectives* 32/3 (2018): 195–214.

20. M. Nikolova and F. Cnossen, 'What Makes Work Meaningful and Why Economists Should Care About It', *Labour Economics* 65 (2020): 101847.

21. Susskind, *A World Without Work*.

22. C. Frey and M. Osborne, 'The Future of Employment: How Susceptible Are Jobs to Computerisation?', *Technological Forecasting and Social Change* 114 (2017): 254–80.

23. D. Autor, 'Why Are There Still So Many Jobs? The History and Future of Workplace Automation', *Journal of Economic Perspectives* 29/3 (2015): 3–30;

L. Tyson and J. Zysman, 'Automation, AI & Work', *Daedalus* (Spring 2022); R. Matheson, 'MIT AI Conference Aims to Prepare Workers for the Jobs of the Future', *MIT News* (22 November 2019).

24. See e.g. Acemoglu, *It's Good Jobs, Stupid*; Acemoglu, 'AI's Future Doesn't Have to be Dystopian';   K. Klinova and A. Korinek, 'AI and Shared Prosperity', *Proceedings of the 2021 AAAI/ACM Conference on AI, Ethics, and Society*, 645–51 (AIES '21, Association for Computing Machinery, 2021).

25. See Susskind, 'Work and Meaning in the Age of AI' for an extensive discussion of these two dimensions.

26. Susskind, 'Work and Meaning in the Age of AI'.

27. J. Cohen, 'Good Jobs', MIT Research Brief (October 2020).

28. For instance, P. Van Parijs, *Basic Income: A Radical Proposal for a Free Society and a Sane Economy* (Harvard University Press, 2017); P. Tcherneva, *The Case for a Job Guarantee* (Polity Press, 2020).

29. D. Susskind, 'A World With Less Work', *Boston Review*, 20 May 2021.

30. Acemoglu, 'Where Do Good Jobs Come From?'

31. Acemoglu, *It's Good Jobs, Stupid*.

32. Acemoglu, 'Where Do Good Jobs Come From?'

33. Acemoglu, *It's Good Jobs, Stupid*.

34. A. Korinek and M. Juelfs, 'Preparing for the (Non-existent?) Future of Work', Brookings Center on Regulation and Markets Working Paper No. 3 (August 2022).

Further Reading

D. Susskind, *A World Without Work: Technology, Automation, and How to Respond* (Allen Lane, 2020).

8

Losing Skills

Carissa Véliz

'Use it or lose it,' my Latin teacher used to say to me, as he handed me homework every day. He was right. Decades later, I wouldn't be able to decline Latin verbs if my tenure depended on it.

I'm writing these words on an airplane. It's cold and windy, and the pilot has mentioned the possibility of turbulence. If there was an emergency, I find myself wondering, does the pilot have enough flying experience to know how to weather it?

On Continental Connection Flight 3407 in 2009, there was no mechanical failure. The captain had been distracted talking with the first officer. As they prepared for landing, they continued chatting, forgetting to monitor the airplane's airspeed and altitude. By the time the captain realized they were in trouble, it was too late. No one on board survived. Similarly, on Asiana Airlines Flight 214 in 2013, a plane crashed because the pilots were not proficient in landing without the use of high-level automation. That day, part of the airport's instrument landing system that helps guide planes to the runway was out of service for repairs.

As flying has become more automated, pilots have been losing certain skills required to fly manually, like navigating by reference to landmarks, calculating their speed and altitude, and being able to visualize the plane's position. They don't get to practise enough. What's more, with automation, they have fewer details to worry about during flights; this induces boredom and mind-wandering, which in turn causes mistakes. When automation fails or necessitates human input, distracted pilots are less able to overcome risky situations.[1]

Artificial intelligence (AI) is the ultimate kind of automation. The aspiration is to create a kind of intelligence that can take over as many of our tasks as possible. As we increasingly rely on AI in more spheres of life—from health and policing, finance to education, and everything in between—it's worth asking ourselves whether increased automation will lead to a loss of expertise, and to what extent that might be a problem.

The concern that technology might degrade our cognitive abilities is hardly new. In Plato's *Phaedrus*, Socrates famously argued that writing would cause people to rely too heavily on external sources rather than on their own memories and understanding:

> [Writing] will atrophy people's memories. Trust in writing will make them remember things by relying on marks made by others, from outside themselves, not on their own inner resources, and so writing will make the things they have learnt disappear from their minds. Your invention is a potion for jogging the memory, not for remembering. You provide your students with the appearance of intelligence, not real intelligence. Because your students will be widely read, though without any contact with a teacher, they will seem to be men of wide knowledge, when they will usually be ignorant.[2]

Socrates thought that writing could not provide the same level of understanding and dialogue that could be achieved from verbal communication, where ideas could be challenged and refined through conversation. In contrast to live dialogue, the written word lacked the ability to adapt to different contexts and therefore could be easily misinterpreted or misunderstood. Writing doesn't talk back:

> The offspring of painting stand there as if alive, but if you ask them a question they maintain an aloof silence. It's the same with written words: you might think they were speaking as if they had some intelligence, but if you want an explanation of any of the things they're saying and you ask them about it, they just go on and on for ever giving the same single piece of information.[3]

Was Socrates right to be worried about writing? I'm torn. On the one hand, I feel fiercely defensive of the written word. Books are one of the best parts of life. They allow you to live many lives in one, travelling to faraway places from the comfort of your living room couch, exploring fresh ideas, meeting new people. Writing kept Socrates alive for all these centuries. Thank you, Plato.

Writing has made it easier to accumulate and share knowledge. Even if our ancestors had prodigious memories, they wouldn't be able to memorize the whole Bodleian Library, to which I'm lucky enough to have access. And sharing knowledge orally is much more inefficient. These days, it is common for people in Mexico to listen to each other's voice messages at twice normal speed to save time. (Yes, they sound bizarre, like cartoon chipmunks on steroids. And if your ear is not accustomed to it, you might not understand a word.) Ancient Greeks didn't have the technical ability to double their talking speed.

Not only does writing facilitate the accumulation and dissemination of knowledge; it also enhances thinking. The slowness of the writing process allows for paced reflection, and the possibility of editing invites fine-tuning. Having words on an external medium like paper or a screen offloads cognitive processing, making it easier to sustain complicated arguments in one's (somewhat external) mind. Just as it is easier to perform a long mathematical computation with pencil and paper than in one's head, so it is easier to handle a complicated argument when you can see the premises and conclusion in front of you. Writing is not only about expressing what is in your mind, but about figuring out what you think as you go along.

However, Socrates was right in that writing has probably damaged our memories. Centuries ago, people used to memorize the equivalent of whole books, word for word. If you were born in the 1980s or before, you probably remember the times when it was normal to memorize dozens of phone numbers. I still remember the phone number of my childhood home, but I have to look at my mobile phone to give someone my current phone number (which I've had for years).

So what? Is memory important? If my phone is always with me, why does it matter if I can't even remember my own number? One key element is reliability. The so-called extended mind theory proposes that our cognitive processes extend beyond our brain and body to include tools, technologies, and the environment we *reliably* interact with.

If you were born before 1960, you are probably much better at arithmetic than younger generations who grew up with calculators. But younger generations don't seem to be missing out because calculators are so easily attainable and reliable. There's a calculator on my phone, so not being great at arithmetic isn't a big handicap. According to the extended mind theory, I have simply offloaded my arithmetic cognitive processes to my phone.

It follows that one important question when it comes to AI is whether it is reliable. If it is reliable, then our losing some skills doesn't seem very alarming. But there are at least four ways in which AI can be unreliable.

First, unlike calculators, AI is (currently) expensive. Running powerful systems takes a lot of computation, which in turns needs a lot of energy. And gadgets like phones and computers which use AI are likewise expensive. Chips, batteries, and devices rely on raw materials like lithium, cobalt, and nickel, and these are finite resources that might eventually run out.

Second, most applications using AI are linked to the internet. Because AI needs considerable computing power and the capacity to store gargantuan amounts of data, it usually attaches to the cloud (to a remote server where heavy computation is carried out). Anything connected to the internet can potentially be hacked.

Ransomware is an example of the risks. Two decades ago, factories, power plants, hospitals, airports, and offices were operated with analogue tools; these were usually more robust than our digital equivalent. Many of today's institutions no longer have the option to operate manually, or have not maintained employees' analogue skills. When metals and electricity company Norsk Hydro received a ransom demand, they managed to avoid shutting down by going manual, but only thanks to older employees and other workers who returned from retirement to help out.[4] Twenty years from now, the generation that knows how to run things on analogue will no longer be around. The task ahead is to make the digital world much more robust than it is currently, whilst at the same time brushing up on our analogue skills as backup.

Third, AI can be unreliable because it is managed by a few powerful tech giants. A company like OpenAI could decide to increase their prices or introduce exploitative terms and conditions, and if we have come to depend on their AI, we will be at their mercy. Alternatively, the company might be sold (as in the case of Twitter, now X) and their algorithm could change for the worse without users having any control over it.

Fourth and most importantly, current AI has an unreliable relationship with the truth. The most popular kind of AI is based on machine learning. An AI like ChatGPT works through statistically analysing texts it has been fed and generating convincing responses based on its training data. But it doesn't use logic and is not tied to empirical evidence. It doesn't have tools to track truth. As a result, it often 'hallucinates', or fabricates plausible responses (based on its statistical analysis) that are nonetheless false. When I asked it to cite ten books written by David Edmonds, it correctly cited *Wittgenstein's Poker* and *The Murder of Professor Schlick* (but got the year of publication wrong on both), it invented that he authored the *Oxford Handbook of Philosophy and Literature* (that book exists, but is edited by Richard Eldridge), and it completely made up a book, *Philosophy 50: Ideas in Brief.* These mistakes are not rare. When I asked it to cite ten books by Carissa Véliz, it invented nine. What is most interesting, and potentially dangerous, is that it came up with plausible titles, given my specialization (e.g. *Understanding Privacy*).

The unreliability of AI should make us think twice about the skills we're losing to it. Even if AI were more reliable—cheap, easily available, unhackable and unpluggable, not in the hands of a few powerful corporations, and trustworthy—we might have reason to hesitate. Recall Socrates, writing, and memory. Even if our new external memories are more reliable than our old internal ones, we might still have lost something important by offloading much of memory onto the written word.

Memory and attention are related.[5] The key to remembering something is being able to attend to it closely. If writing has weakened our memories, perhaps our attention spans have suffered too—and maybe attention will be further eroded with the development of digital tech.

As I write this chapter, my attention keeps jumping around—from the blank page to some references, from an annoyingly addictive game on my phone to a text message from my mum, from social media notifications to online newspapers, and half an hour later back to the blinking cursor on this page—all the while holding me back from the pleasurable state of flow that can kick in after hours of sustained concentration. It is possible that the attention span of Ancient Greeks was far superior to ours, and with it came pleasurable experiences such as being in the moment and experiencing flow states more frequently than we do.

There is often an assumption that greater automation will give us the ability to focus on higher-level tasks. That seems unlikely. Thanks to the washing machine, I might not spend as much time washing clothes as my grandmother did, but I spend more time on email than she ever did washing clothes. And I bet she had more high-level thoughts while washing clothes than I'm able to have while emailing. The experience of washing clothes—the feel of the water running through your hands, for example—is also more pleasant for our embodied experience than staring at a screen.

When it comes to writing, automation might weaken or wipe out some of the critical thinking skills that we gained when we chose to ignore Socrates and bargained away our memories. If we use chatbots to write for us, in a best-case scenario, we might get an acceptable product with a fraction of the time and effort that would've been involved in writing manually, but we will have missed out on the process. Writing—the process, not the product—is a cognitive enhancer. It is a kind of supercharged thinking.

Among the abilities that writing improves is empathy. Writing asks you to put yourself in the reader's shoes. Will they understand what you are trying to communicate? Are you sure you won't bore them to death? Are you being respectful enough to other points of view? Social skills are some of the most valuable skills we possess; we would do well to make sure they are not lost to AI.

If our students become used to relying on chatbots to write their essays, they might lose critical thinking skills, empathy, and creativity. Their essays might become (even more!) full of common tropes and cliched observations. As AI retrains on the language it produces (as opposed to human-made texts), it might become more hallucinogenic.

Could ChatGPT produce an anecdote like the one with which this chapter began—of my Latin teacher and how we can't retain what we don't exercise? Let's see. ChatGPT, come up with a short, illustrative, and credible anecdote of how we lose the abilities that we don't use:

> A woman used to be an excellent painter, but she stopped painting when she got a full-time job. Years went by, and when she tried to paint again, she found that her once-impressive skills had deteriorated.

A bit bland if you ask me, especially in the use of language. Then again, as the airplane I'm on lands smoothly on the runway, it occurs to me that maybe you think that my Latin declensions are not that interesting either. Touché. Be that as it may, you should still pay heed to my former teacher—and use the skills that you don't want to lose.

Notes

1. <https://www.newyorker.com/science/maria-konnikova/hazards-automation>.
2. Plato, *Phaedrus*, ed. Robin Waterfield (Oxford World's Classics, 2009), 275a.
3. Plato, *Phaedrus*, 275d.
4. <https://www.ft.com/content/c4ea6fb3-6262-4426-9503-05391f0e523a>.
5. <https://www.ncbi.nlm.nih.gov/pmc/articles/PMC6688548/>.

Further Reading

Carissa Véliz (ed.), *Oxford Handbook of Digital Ethics* (Oxford University Press, 2021).

9

Benevolent Algorithmic Managers

Charlotte Unruh

Imagine a company that uses an algorithmic system to manage their workforce. The recruitment system selects applicants that are a good fit, the task allocation system matches workers to suitable tasks, and the team allocation system assigns workers to teams that work productively. By increasing worker satisfaction and motivation, we might imagine, this system also benefits the company's bottom line.

But does this seemingly attractive workplace come at a cost?

The Smart Company

On Monday morning at 9:50, precisely the time set by the Smart Company's shift allocation system, Billy enters the Smart Company's building. A quick glance on his device tells Billy that he is checked in for his shift, and notifications on his screen guide him to a workstation next to the storage area. Once there, Billy picks up some items waiting to be wrapped, and then takes another look at his device to check his instructions for the day. The Smart Company's task allocation system matches workers with tasks for which they are well suited. Billy wraps some items for transport and then continues working through his tasks, guided by and aiming to fulfil the targets set by the system. After lunch, Billy attends several meetings. The team allocation system has assigned Billy to teams that will work together productively.

The system has been in place for several months. Today, for the first time, it occurred to Billy that he had no idea how the system operated—and he asked his boss. 'It's far too complicated to explain,' came the answer, 'but don't worry, there's nothing wrong with the algorithm. The tasks you're given will be fun and fulfilling, the allocation of work will be fair, and teams will collaborate effectively.' And indeed, Billy has up until now had no complaints about the system's decisions.

Is there anything morally problematic about the Smart Company's algorithmic system? In this chapter, I argue that the system restricts the autonomy of workers. This is a morally relevant consideration, whether or not Billy is happy with its decisions.

My argument rests on the idea that autonomy is a fundamental moral value. As humans, we govern our own lives: we form beliefs and attitudes, we act based on our beliefs and attitudes, and we hold others to account when they interfere with or undermine our actions. By regulating every aspect of Billy's work day, the Smart Company's algorithmic system is in tension with Billy's autonomy, and specifically, his ability to exercise discretion in his work.

Algorithms at Work

The Smart Company case draws inspiration from real-world developments. Algorithmic systems are increasingly part of our working lives. Algorithms can support, and potentially replace, human decision making in areas such as recruitment, scheduling, performance monitoring, task allocation, and team formation. In this way, algorithms take on the role of managers in monitoring and directing human workers.

What is new about algorithmic decision making in the context of work? Algorithmic systems rely on vast amounts of data that human managers could not collect or process. The large-scale collection of data is often only possible through surveillance and tracking of worker behaviour—for example, sensors in warehouses, tags embedded in products, and devices worn or carried by workers. Based on this data, algorithmic systems can fulfil different functions. The data collected by sensors, devices, and machines can be used to *analyse* processes and to make predictions about future processes. For example, data analysis might reveal patterns of demand or resource use. Algorithmic systems might also use past data to *predict* outcomes, for example the performance of specific employees. And algorithmic systems can make *recommendations*, for example regarding the composition of teams based on data on past team performance. The availability of data enables the far-reaching optimization of industrial processes. For example, algorithms might analyse the routes that workers take through warehouses and calculate more efficient routes.

Moreover, algorithmic systems can exert fine-grained control over workers. Algorithmic systems can reduce workers' *discretion* in how they do their tasks. Workers in warehouses can, through their devices, receive detailed instructions regarding where to go next (to ensure the most efficient route), and which item to take off the shelves. Control can be exercised directly, for example by directing workers via wristbands that vibrate when workers make mistakes, or more indirectly, for example by using algorithmic nudges and gamification. Amazon reportedly uses video games to let warehouse workers earn points and badges and compete with each other by completing work tasks.[1] In these ways, algorithms enable new forms of employer control. Employers can direct, evaluate, and discipline workers in a much more far-reaching way than previously possible. Algorithmic systems can also change the nature of the task themselves. Algorithmic management facilitates the *fragmentation of tasks*—breaking tasks down in small chunks of work that can be centrally monitored and coordinated. Digital interfaces can provide guidance and direction in performing these small steps, making the work easier to perform, including by untrained workers, but leading to micromanagement that might leave workers with little discretion.

The Argument from Sovereignty

The philosopher Andrea Veltman[2] helpfully distinguishes three notions of autonomy in relation to work. First, autonomous work can be defined as work that is *autonomously chosen*. Second, autonomous work could mean that work provides *economic independence* by providing a salary and other benefits. Third, autonomous work can mean that the worker has *autonomous agency* in their work. Here, I am interested in autonomy in this third sense. Workers who are autonomous can make their own decisions about, for example, the kind and order of tasks, the way in which to go about doing these tasks, and the time and place of work. On this understanding, autonomy in work is about having discretion in decision making.

In philosophical terms, this path prioritizes *sovereignty*[3] or *independence autonomy*.[4] Consider Horacio Spector's definition of sovereignty.

Sovereignty: 'If S does φ, then S's φ-ing is caused by S's choice to φ.'[5]

According to this definition, whether an agent is sovereign depends on whether the agent's choices cause their actions. Imagine an ice cream van that, for promotion purposes, hands out one and only one scoop of ice cream to each passer-by. Annie chooses between different flavours of ice cream: chocolate, vanilla, and strawberry. Annie is sovereign with regard to her choice of ice cream when Annie's eating strawberry ice cream is caused by Annie's choice to eat strawberry ice cream.

Here are two ways in which sovereignty can be violated. First, Annie might not choose to eat strawberry ice cream. For example, if Bob forces Annie to eat strawberry ice cream, then Annie is not sovereign in her choice. Second, according to the definition above, what matters is not just that Annie eats strawberry ice cream and that Annie has chosen to eat strawberry ice cream; there must also be the right kind of connection between Annie's choice and Annie's action. The choice must cause the action. But this causal connection can be severed. For example, imagine that Annie chooses strawberry ice cream. The ice cream vendor thoughtlessly gets a scoop out of the nearest tub. Luckily for Annie, it is strawberry ice cream. But this is pure coincidence. Annie's choice does not cause Annie's consumption of strawberry ice cream. Now imagine that the ice cream vendor watches Annie as she approaches. Because she is wearing a blue dress, and because, in the ice cream vendor's experience, most women in blue dresses order strawberry ice cream, he hands her a strawberry ice cream before she has a chance to request it. Again, Annie's choice has not caused her to eat strawberry ice cream.

Let us now return to Billy in the Smart Company. Billy's working on a specific task is not the result of Billy's choice. Billy was assigned this task by the algorithmic system. In this sense, Billy's action was not sovereign. The algorithmic system makes decisions about the tasks that Billy is allocated, the pace and timing of work steps, the teams in which Billy works, and so on. In these ways, the algorithmic system restricts Billy's sovereignty.

This is the case even if Billy is happy with the tasks assigned to him. The algorithmic system steps into the causal chain between Billy's choice and Billy's action, in a similar way in which the ice cream vendor steps into the causal chain between Annie's choice and Annie's action. Since the algorithmic system assigns Billy's tasks, his choices do not cause his actions, even if they happen to align with them. But as the literature on autonomy highlights, choosing itself has value.[6]

It is important to note that the argument that work restricts the autonomy of workers is not new. Social and political philosophers have long argued that corporations problematically undermine the autonomy of human workers by subordinating their interests and needs to the economic needs of the corporation. So-called 'Taylorist' management divides labour into separate tasks that can be centrally controlled, in an effort to increase productivity. The division of labour and the drive for efficiency that permeates offices, warehouses, and factories reduces worker control over their daily tasks, broader work environment, and strategic decisions at a company level.[7]

But while the threat to the sovereignty of workers has existed for some time, algorithmic management makes the threat more salient, and more pronounced. For example, the fragmentation of tasks might simplify work and make it less varied, and autonomy regarding the time and place of work might blur boundaries between work and home, with the result that workers feel compelled to check emails or do work tasks outside working hours and away from the office.

The sovereignty of workers matters, even if algorithmic systems successfully adapt work to worker preferences.[8] To make clearer why benign systems can still violate sovereignty, it might help to consider a case given by David Enoch:

The benevolent pollster: 'The benevolent pollster is a dictator [. . .] [who] routinely conducts polls to find out information about the values and deep commitments of those he rules over. He does this not out of some cynical motive (say, to increase his chances of retaining power), but out of a deep moral commitment—perhaps, for instance, he accepts the value of autonomy as non-alienation, so it's very important for him to make sure, to the extent possible, that the life-stories of his subjects are in line with their values and deep commitments. This is why he designs his policies in ways that harmonize with the results of his polls.'[9]

Enoch argues that the citizens in this case are not autonomous in the sovereignty sense. The benevolent pollster governs a dictatorship, not a democracy. Perhaps we cannot straightforwardly draw analogies between corporations and countries on the one hand, and workers and citizens on the other. However, the case of the benevolent pollster illustrates how even a benevolent algorithmic manager is in a position of power vis-à-vis workers. This gives us a prima facie reason to be wary of benevolent algorithmic management.

The Objection from Non-Alienation

In this section, I consider an objection to my argument in the last section. The objection builds on a different notion of autonomy. The actions of people who are autonomous in this other sense reflect their deep commitments and values. This aspect of autonomy is *non-alienation*[10] or *real-self autonomy*.[11] Here is Horacio Spector's definition:

Non-Alienation: 'If S chooses to do φ, then S's φ-choosing reflects S's real self'.[12]

When we understand autonomy as non-alienation, the Smart Company case might seem less problematic. After all, we have stipulated that the algorithm works well: it assigns Billy tasks he enjoys, co-workers he likes, and so on. This observation might lead one to argue that Billy's preferences reflect his 'real self', his commitments and values. If the Smart Company ensures that the tasks workers perform match their preferences more closely than before, the Smart Company actually enhances the non-alienation autonomy of workers. The infringement of Billy's sovereignty might be outweighed by the promotion of Billy's non-alienation autonomy.

In response, note that the objection presupposes that what matters morally is the *overall* impact of the algorithm on Billy's autonomy. But perhaps this is not all that matters. Perhaps sovereignty matters morally, independent of Billy's overall autonomy. Or sovereignty matters more, or in a different way, from autonomy as non-alienation. For example, Enoch[13] claims that while non-alienation is the more fundamental value, sovereignty is the more central value in political contexts. The objector would need to show that gains in non-alienation should count as much in favour of the Smart Company's use of its algorithmic system as losses in sovereignty count against it.

Another reply to the objection is to reject the view that the Smart Company's algorithm increases Billy's autonomy in the non-alienation sense. After all, it seems odd to say that workers' preferences of the sort that can be satisfied by the Smart Company reflect the worker's deep commitments or values. A deep commitment might be something like a passion for teaching, or care for one's family. I might prefer to start my workday with easier tasks, or I might take more breaks in the afternoon than in the morning, but it seems a stretch to say that these preferences express deep commitments and values.

Perhaps, this reply is too strong. For surely, preferences can be grounded in deeper commitments. For example, a supermarket worker might see social relationships as an important part of their identity and enjoy interacting with customers rather than stacking boxes in the storage space. So a better way to put the reply is to say that not just any choice that a worker is happy with, or that aligns with their preferences, increases their autonomy in the non-alienation sense.

Moreover, there are ways in which sovereignty might be closely connected to our real self. Some people might have a preference to exercise workplace discretion, and this might be related to their higher-order commitments, for example their commitment to leading a self-determined life or to exercising their decision-making capacities. Task fragmentation and micro-management might be in tension with the higher-order commitments of such people.

None of this shows that we should not aim to ensure that workers have tasks that they prefer. Doing so might be very valuable, insofar as it is likely to increase their wellbeing. My point is that aligning work to worker's preferences, in itself, neither necessarily increases worker autonomy in the non-alienation sense nor their sovereignty.

Conclusion

We are now in a position to see what might be morally problematic with the Smart Company. The algorithmic system controls Billy's working life. It decides what tasks Billy does, who he works with, and how his work is done. In doing so, the system limits Billy's sovereignty. This remains the case even if the system makes good choices from Billy's point of view.

I do not claim that limiting the sovereignty of workers is always problematic. Perhaps hierarchical institutions like workplaces only function when the sovereignty of workers is limited. Nor do I claim that technology should never limit worker autonomy. Delegating decisions to algorithms might have advantages. For example, some algorithmic systems might be more transparent and fairer than some human managers.

My point is that there is value in sovereignty. Lack of discretion at the workplace threatens sovereignty, and algorithmic management can make this threat more pronounced. We should consider impacts on worker sovereignty when deciding whether to use algorithmic systems at work and how to design them. Perhaps we can align work with worker preferences in ways that do not limit sovereignty. For example, involving Billy in decisions might enable him to shape his work to his needs and preferences without limiting his sovereignty. If they can facilitate worker participation in work design, algorithmic tools might even help enhance sovereignty. In one study[14] a participatory method was used for eliciting workers' scheduling preferences. The researchers provided a web tool where workers could provide information about their preferences, such as their preferred shift start time and whether or not to work weekends. The web tool then created a model representing the worker's preferences. Workers could then adapt the model to ensure that their preferences were adequately represented.

These or similar methods offer hope for the future world of work: that workers might gain greater control over their tasks and schedules and so increase their sovereignty.

Acknowledgements

I thank David Edmonds for excellent feedback, and Charlotte Haid, Johannes Fottner, and Tim Buthe for helpful discussion.

Notes

1. Greg Bensinger, '"MissionRacer": How Amazon Turned the Tedium of Warehouse Work into a Game', *Washington Post*, 24 May 2019.
2. Andrea Veltman, *Meaningful Work* (Oxford University Press, 2016), 77–85.
3. David Enoch, 'Autonomy as Non-Alienation, Autonomy as Sovereignty, and Politics', *Journal of Political Philosophy* 30/2 (2022): 143–65.
4. Horacio Spector, 'Autonomy and Rights', in Ben Colburn (ed.), *The Routledge Handbook of Autonomy* (Routledge, 2022), 313–23.
5. Spector, 'Autonomy', 315.
6. This point is made by Gal in the context of algorithmic assistants, who notes that '[w]hile we may have many options open to us, the choice among them is not made by the individual based on self-reflection and self-deliberation that lead to the shaping and application of his preferences, but rather through an algorithm which is assumed to be able to mimic these preferences' (Michal Gal, 'Algorithmic Challenges to Autonomous Choice', *Mich. Tech. L. Rev.* 25

(2018): 89). See also Joseph Raz, *The Morality of Freedom* (Oxford University Press, 1988), who emphasizes the value of exercising (vs possessing the capacity for) autonomy, and John Danaher, *Automation and Utopia* (Harvard University Press, 2019: 125) who argues that automation undermines agency.

7. Breen argues that the division of work violates the conditions for the autonomy of workers, and moreover, 'hinders the maintenance of a healthy practical identity, insofar as it impedes rather than augments most reasonable life plans' (Keith Breen, 'Meaningful Work and Freedom: Self-Realization, Autonomy, and Non-Domination in Work', in Ruth Yeoman, Catherine Bailey, Adrian Madden, and Marc Thompson (eds), *The Oxford Handbook of Meaningful Work* (Oxford University Press, 2019), 60).

8. This point is also made by Gal ('Algorithmic Challenges to Autonomous Choice', 89), who points out that perfectly benign algorithmic systems might be even more problematic than less benign ones as we would be more likely to overly rely on such algorithms.

9. Enoch, 'Autonomy'.

10. Enoch, 'Autonomy'.

11. Spector, 'Autonomy'.

12. Spector, 'Autonomy', 315.

13. Enoch, 'Autonomy'.

14. Min Kyung Lee, Ishan Nigam, Angie Zhang, Joel Afriyie, Zhizhen Qin, and Sicun Gao, 'Participatory Algorithmic Management: Elicitation Methods for Worker Well-Being Models', in *Proceedings of the 2021 AAAI/ACM Conference on AI, Ethics, and Society* (AIES '21: Association for Computing Machinery, 2021), 715–26.

Further Reading

Karen Levy, *Data Driven: Truckers, Technology, and the New Workplace Surveillance* (Princeton University Press, 2022).

10
What's Your Price?

Three Cheers for Data Markets

Aksel Sterri

Introduction

The great thing about the online world is that there are no class divisions. Everyone, regardless of their wealth, success, and status, can access many of the most valuable online services, and we are free to come and go as we want. This stands in sharp contrast to the offline world, where everything has a price, scarce goods go to the people who can afford them, and workers are under the thumb of some master.

This idyllic picture of the online world conceals, however, that free and cheap services are only possible because the data we produce are taken and exploited for financial gain by powerful tech companies. And below the surface is a world of manipulation, mass surveillance, and misaligned incentives that lead to socially harmful outcomes.

To correct these problems and reach a more honest online world, we must pay for other people's data and get paid for our own. We need data ownership, data markets, and data cooperatives that provide people with effective control.

The Current Shape of the Digital Economy

Whenever we do something online or interact with a computer in the physical world—by swiping our credit cards or driving a car—we leave a data trail. This data is used to understand who we are, what makes us tick, what we are likely to do, and how cars and other machines operate. Data, simply understood, is all information that can be read by a computer. One's date of birth, hair and eye colour, genome, location, preferences, friends and acquaintances, personality, and intellectual and artistic contributions are either data or information that can be transformed into data.

Data is big business. 23andMe interprets your genetic data for you in exchange for a fee and the right to keep and use your data. Amazon and Spotify use data about their users to offer tailored recommendations for books and music. Alphabet (Google) and Meta (Facebook and Instagram) use data about their users to sell their attention and readiness to pay for goods and services to companies that want to sell stuff online. OpenAI, Alphabet, and Meta use enormous amounts of data that they scrape off the web to train large language models, like GPT-4. They aim to create increasingly more sophisticated artificial intelligence that can replace human labour for an increasing number of tasks.

The current shape of the digital economy is radically opposed to several of our core interests. The first casualty is our privacy interests. The digital economy requires mass surveillance in order to function. Our economic interests are also on the line. Tech companies are very profitable but produce very few jobs. And the development of ever-smarter artificial intelligence may eventually make the very workers who provided the data necessary to train the AI redundant. Finally, the digital economy alienates us and threatens our autonomy. Services like Facebook, Instagram, and X (previously Twitter) are designed to manipulate their users to act in ways that are beneficial to advertisers, not to serve the interests of users themselves or the wider society. The simple explanation for this is that money is power: companies do the bidding of those who pay.

The Tripartite Model for Data Control

To do better, we need three reforms that together make up a tripartite model for data control.[1] The first is to legally recognize that ordinary people, as *data producers*, have a rightful claim to control the data they produce. This differs crucially from the current system, which operates according to a rule of capture: whoever records the data gets to keep it and reap all the benefits from its use.

The second is to implement two data markets. One is a *labour market* in data, where tech companies and other data buyers can request data from data workers in exchange for payment. The other is a *leasing market* in data, where companies pay data producers for the right to use data in specific contexts.

The third is the creation of cooperatives that serve as intermediaries between data buyers and data workers. These co-ops will represent the interests of their members in negotiations with companies and government agencies that want to use data, as well as with other co-ops in case of conflict over data; they can also serve an educational function for their members.

Benefits of the Tripartite Model

The tripartite model has several benefits. Let us begin by focusing on the benefits to data producers. By facilitating a labour market and the leasing of data for specific purposes, people gain control over who gets access to what data and for what purpose. These markets will therefore protect privacy.

The model will also promote people's economic interests. As people make money from the data they produce, they get a stake in the booming digital economy. If companies that train smarter algorithms become rich, some of that wealth will accrue to everyone who contributes with data.

The model may promote people's interest in shaping society. Currently, data is used by anyone who can access it. But if data producers have control rights, they can exclusively labour for and sell user rights to companies they believe work for sufficiently good causes. Moreover, when data production is recognized and rewarded, and people have more control over their data, they may feel less alienated from the increasingly digital society of which they are a part.

Objections

Do Companies Have a Stronger Claim to the Data?

One objection to giving control rights to data producers is that companies have done more to produce data, and therefore have a stronger claim to it, than anyone else. We tend to think that when someone is creating a piece of art using their talent, ingenuity, and effort, it rightfully belongs to them. But when people use Facebook, they do not intentionally seek to create data, nor would this particular set of data have been produced without Facebook's infrastructure. If one acquires ownership rights because of effort and creativity, companies may have a greater claim to the data than people who click on a link to purchase a pair of socks or watch cat videos.

One response is that this may be true for some data but not for all data—for example, the data that goes into training large language models. We should certainly grant that transforming behavioural patterns on websites into data, and to make online text, pictures, and video readable for algorithms, takes hard work from employees in OpenAI, as well as considerable ingenuity. But more important is the human labour and creativity that go into producing everything that is now available online. Without this, these models could never perform their miracles.

A more fundamental response is to deny that it is appropriate to ground data property rights in effort and creativity. If labour and creativity were sufficient to give rights to data, one could acquire rights to control other people's personal information. For example, when I post pictures of my friends on Instagram, I use my labour to create data about where my friends are and who they are with. It is more plausible that sensitive and private information belongs to the person to whom it is about. Without control over data that reveals information about us, privacy would be impossible.

The importance of safeguarding people's privacy is part of an altogether different justification for granting control rights to data producers—that it will secure more of the things we value.

As we saw above, the benefits involve improving people's economic interests, autonomy, and power.

Does the Tripartite Model Prevent the Flow of Data?

At this point, one might reasonably object that while data producers may benefit *as producers* they may be worse off overall. The new legal framework might prevent data from being put to its most beneficial use.

This objection correctly points out that we must assess the overall effects of giving control rights to data producers. It also points to the importance of letting data flow. Data has the rare property of being *non-rivalrous*. It is like a large ocean in which one person's use of the ocean doesn't encroach on its use by others. That data can be used by many people at once is one of its great sources of value and why it is particularly detrimental if data is prevented from flowing to all who can gain from it.

However, giving control rights to data producers is likely to enhance, not prevent, the flow of data compared to the existing order. Data is information, and information gives companies a competitive advantage. Companies therefore have a reason to prevent competitors from accessing data. Data producers, on the other hand, will have an economic incentive to sell user rights to data to as many competitors as possible, provided their privacy interests are secured and companies are operating in ways aligned with their views on just and decent conduct.

Incentives

Companies may not have a moral right to control the data they co-produce with the users for the reasons presented above, but they are profit-maximizing entities. If they do not have data rights, they might not be financially motivated to create the necessary infrastructure for producing data. So, perversely, giving control rights to people who are data producers may undermine their economic interests.

One response is that a data market will also benefit companies. It incentivizes people to produce data of better quality since companies will be willing to pay more for high-quality data. For example, if Facebook could ask its users to give accurate captions on uploaded pictures, its machine learning algorithms would become more accurate. High-quality data could make companies more effective and profitable; they would no longer have to tease out the signal from the noise in messy surveillance data.

A second response is that co-ops could make agreements with companies on a case-by-case basis, where companies that are essential to producing data are granted either a share of the income of future data leases or exclusive use for a limited time period. This latter option is equivalent to incentivizing drug development by giving pharmaceutical companies the exclusive right to sell their new drugs for a period of time.

Privacy

A concern with the tripartite model is that data producers would not protect their privacy interests. Currently, consumers show very little concern for their privacy. They continue to use services from companies that are widely known to spy on their users. And when asked about how much they are willing to pay to protect their privacy, the numbers tend to be small. To the extent that data ownership and data markets are supposed to protect people's privacy interests, this could be a reason to seek other solutions.

However, we need not accept the premise that people don't care about their privacy. A plausible explanation for people doing little to protect their privacy is that they are rationally apathetic. Even if they think that privacy is important, they realize that it is not something they can secure on their own.

Indeed, if a company has access to data about people like me, they can also infer much about me, whether or not I have gone to any trouble to protect my privacy. This gives rise to a 'tragedy of the commons' situation, where what is in the interest of each person acting in isolation is not what is in the collective interest.

If the value an individual person gets from protecting their privacy is small unless other people also protect their privacy, they will understandably not go to great lengths to secure it. This gives rise to a 'tragedy of the commons' situation, where what is in the interest of each person acting in isolation is not what is in the collective interest. However, this problem can be mitigated if people partake in data co-ops. By collectively deciding how much concern they should give to privacy, they can protect their common interest in privacy.

If people in this situation still choose to trade some privacy for other benefits, this may be a strength, not a weakness, of the model. Some interests, like those that are protected by civil and political rights, are too important to sell. But privacy is not such a core and inalienable right. It can be reasonable to share even very personal information if the benefits from doing so are sufficiently large.

What is deemed reasonable might depend on how much money you have and how much you are set to earn by selling your data. This raises the concern that only well-off people will protect their privacy since money is worth more to those with little of it. This is certainly a problem, but the solution is to increase the income for poor people, through cash transfers or job opportunities, not prevent people from engaging in markets that allow them to make trade-offs between money and privacy.

Can I Sell Your Information?

If one person's data provides information about other people we may wonder whether they should be free to sell it. The most obvious instance of the relational nature of data is that my genetic data contains information about my family's genetic material. The point can be made more generally: given that I am like people who are similarly situated, my data reveals information about them. And this may threaten their interests.

That we have overlapping datasets is another reason why we need data co-ops. These institutions facilitate forums in which people can collectively agree on what data to sell and not to sell instead of relying on data markets in which individuals operate alone. Joining forces through a data co-op will also strengthen data workers' bargaining position vis-à-vis companies. When data is relational, there is a risk that individual data sellers will undermine the bargaining position of others. My data is less valuable if a data buyer can get much of its value from others. If people with overlapping data could join together, they could demand a higher price than they would be able to achieve if they had to compete against each other.

Is Free Better?

Another objection is that the tripartite proposal will be detrimental overall because our data is less valuable than we assume, and the benefits of free services are greater than we admit. One way to estimate the value of the services you currently receive is to ask yourself how much you would require to be barred from accessing services like Google Search and Google Maps, as well as ChatGPT and other AI tools. Compare this sum to your best estimate of how much an average person can expect to get in compensation for the data they produce. If the latter sum is less than the former, we would be worse off.

There is no point in denying that we currently receive many valuable services for free or at a subsidized price. But we should deny that our data is of very little value. The going rate for data in an unregulated market is neither the only nor the best way to assess the value of data. Indeed, since each of us produces data that overlaps with data produced by others, the going rate for data is unlikely to correspond to its true value. If we can sell our data collectively, we will receive more than we would alone.

How much is data worth? In *Radical Markets* (2018), Eric Posner and Glen Weyl estimate that if we recognized data labour an American household of four would, within twenty years, increase their income by $20,000.[2]

Inequalities

A related objection is that 'free' is better because it levels our ability to consume. People vary in their ability to contribute to data production but less in their ability to consume. Free is, therefore, more egalitarian.

This is true in one sense, but our *equality as consumers* in the online world conceals other inequalities. The world of free services prevents many new jobs from being created in the online world. If people had to be paid to produce data, they could specialize in providing different types of data relevant to improving artificial intelligence. There is also a substantial risk that increasingly sophisticated algorithms will replace human workers even when this is socially suboptimal. As long as the labour that goes into creating clever algorithms is unrecognized and unrewarded, the true cost of producing and operating them will be obscured. If we replace compensated labour in the offline world with uncompensated labour in the online world, this is likely to be socially suboptimal and exacerbate inequalities. If OpenAI had to compensate all the people who produced data that have gone into training large language models, this powerful technology would not have developed as quickly, nor would it have such disruptive consequences.

Equality as consumers is also a poor substitute for equality as producers. We need to feel recognized and perceive ourselves as value creators. An equality worth pursuing is equality of self-esteem and dignity, not merely equality to use or buy.

Conclusion

A big problem in the digital economy is that data producers are not recognized for their contributions. As long as our role in digital value creation is obfuscated, our power to affect how our data is used will be limited, and we will receive very little of the value that is created. I have argued for three reforms: control rights to data producers, data markets, and data co-ops. This tripartite model would ensure a fairer distribution of power and resources in the digital sphere. It would also secure ordinary people's dignity and self-esteem by recognizing their role as value creators, not merely as beneficiaries of Silicon Valley's technological prowess.

Acknowledgements

I'm grateful for helpful feedback from Teodora Groza, Albert Didriksen, Sadie Regmi, and David Edmonds. The research is funded by the Research Council of Norway, project 315957. I also benefited from conversations with Glen Weyl.

Notes

1. The tripartite model aligns with previous proposals, most notably J. Lanier, *Who Owns the Future?* (Simon & Schuster, 2013); E. Posner and G. Weyl, *Radical Markets* (Princeton University Press, 2018); and RadicalXChange, *Data Freedom Act* (RadicalXChange Foundation, 2020). The idea of a co-op (in the form of 'data altruism' organizations) is already recognized in the European Union's Data Governance Act.
2. Their assumptions are that artificial intelligence will grow to contribute up to 10 per cent of GDP over the next twenty years, that recognizing data labour will increase data workers' share of the income generated in the digital economy to 70 per cent, and that this will increase productivity by 30 per cent.

Further Reading

Jaron Lanier, *Who Owns the Future?* (Simon & Schuster, 2013).

11

Work and Play in the Shadow of AI

John Tasioulas

Introduction

AI promises to be a means of enabling us to perform more efficiently a whole range of activities that standardly have required the exercise of human cognitive powers. As such, it poses a distinctive challenge. What does it mean to find fulfilment as a human being in a world in which AI-based systems are undertaking tasks that have characteristically given human life its point? Many activities give life meaning, from appreciation of the arts to friendship and politics; but I am going to focus on work.

What is work? In a helpful discussion, Raymond Geuss has identified three of its core features: (a) *exertion*—work requires the expenditure of energy and is strenuous; (b) work is a *necessity* of life; and (c) it has an external product that can be measured and evaluated independently of how it came to be (*objectivity*).[1] I am going to follow this characterisation, which has the welcome consequence that the myriad forms of often unpaid domestic labour count as work.[2]

Some might object that the challenge posed by AI technology for work is nothing new, since technological advances have long been a cause of unemployment. But this underestimates the profundity of AI's implications for the place of work in a flourishing life. There has never before been a technology that has the potential to replace human work activities on anything like a comparable scale. This goes well beyond carrying out 'routine' or 'mechanical tasks' to include white-collar occupations, such as journalism and legal services. A 2017 Oxford-based study concluded that 47% of all occupations in the United States are capable of being 'computerised' in the next

10–20 years.[3] Meanwhile, a McKinsey study found that, although just under 5% of occupations are fully automatable, around 30% of all work tasks in 60% of occupations could be automated.[4] Nor can we confidently assume that, as with previous technological advances, new and unanticipated occupations will emerge to replace those eliminated by AI technology.

Well-Being and the Value of Work

Discussions of the impact of AI on human fulfilment or well-being must begin by acknowledging that the latter is a highly contested notion. Three broad types of theories of well-being have come to prominence in philosophy.[5] Hedonism, a view associated with the 18th-century British philosopher Jeremy Bentham, states that the only thing that ultimately makes people better off is pleasure and the absence of pain. Desire-fulfilment theories, widely favoured by contemporary economists, hold that the goodness of our lives is a matter of the fulfilment of our desires or preferences, quite independently of whether this gives us pleasure. Finally, objective list theories identify a plurality of values, the actualisation of which in our lives makes us better off quite independently of whether we desire those values or get pleasure from them. Standard items on such an 'objective list' include knowledge, friendship, achievement, play, and aesthetic experience. In this chapter, I am going to assume the correctness of a theory of this third sort.[6]

A defining feature of modernity is what the philosopher Charles Taylor has called 'the affirmation of ordinary life'. By 'ordinary life', Taylor means 'those aspects of human life concerned with production and reproduction, that is, labour, the making of the things needed for life, and our life as sexual beings, including marriage and the family'.[7] Work is essential to ordinary life. And one of the distinguishing features of modernity is that ordinary life, including work activities, is treated as a scene of genuine human fulfilment. This contrasts with those ancient and medieval modes of thought that tended to disparage ordinary life and saw true human fulfilment as accessible only to the sage or the saint, figures who have liberated themselves from the mundane concern with the production of the necessities of life.

So understood, work is not merely valuable instrumentally, as a means to an income. If its value were purely instrumental, the loss of work opportunities could be fully made good by the establishment of a Universal Basic Income. Rather, the idea is that there is intrinsic value to work. On an objective-list approach to well-being, it is reasonable to suppose that work enables the realisation of a plurality of intrinsic values. Anca Gheaus and Lisa Herzog have identified these as attaining various types of excellence; making a social contribution; experiencing community; and gaining social recognition.[8] Joshua Cohen has highlighted the values of worker voice (i.e. a say in what is produced and how), realising a valuable purpose, and taking pleasure in the work itself, in addition to the 'standard' instrumental goods of jobs such as money, stability, health and safety, etc.[9]

These accounts seem plausible, but they prompt the question whether the intrinsic goods of work comprise an unstructured list, or whether there is a core good, one that is the primary good distinctively associated with work and in relation to which other intrinsic goods of work are derivative or secondary. A plausible candidate for the core intrinsic value is achievement, that is, the exercise of our human powers in meeting difficult challenges for a worthwhile end. And this intrinsic value is increased the more difficult the challenges and the more worthwhile the end. Part of the value of community or friendship in work is cooperating with others in order to realise a challenging but worthwhile task. Moreover, an important spin-off of achievement is the self-esteem that it promotes, which is a significant factor in fostering the sense among citizens of a democratic community that they can look each other in the eye as equals.

Of course, in claiming that achievement is the core value of work, it does not follow that all work activities offer adequate opportunities for the exercise of our powers in engaging in challenging but worthwhile tasks. Some jobs may be mind-numbingly repetitive, and their value to those who do them may just consist in the wages they are paid. Moreover, even if some jobs, like bomb disposal or refuse collection, offer opportunities for achievement, they may be so dangerous or disagreeable that we should be willing to forego those opportunities if those jobs could be done by a robot. In general, however, I think we should be wary of sweeping claims by academic writers that certain jobs, typically of the sort that do not require a university degree, afford little scope for achievement. Often the problem with these jobs is not the absence of opportunity for achievement, but appalling conditions of work, including the low salaries given to people otherwise described as 'essential workers'.

One radical challenge posed by AI is that it forces us to inquire into the sources of human achievement in an age of automation. This challenge arises for one of two reasons. Either there will be too few jobs for humans to do, since they will have been taken over by AI. Alternatively, even if we preserve human work opportunities by deliberately restricting the deployment of AI, those performing the jobs may be haunted by a sense of pointlessness. They will know that what they are doing could be done just as effectively, and a lot more efficiently, by AI technology. This is akin to the sense of pointlessness that assailed those who performed unnecessary government-created jobs in the Soviet Union or who today undertake the sorts of pointless white-collar jobs that the anthropologist David Graeber referred to as 'bullshit jobs'.[10]

In an essay entitled 'Economic Possibilities for our Grandchildren', published in 1930, John Maynard Keynes greeted the prospect of a jobless future with cautious optimism, signalling that it may necessitate a radical but unforeseeable revision of our ethical outlook—'we have been trained too long to strive and not to enjoy', he said.[11] Maybe so. But still, on an objective list approach, passive enjoyment cannot fully displace achievement as a dimension of the human good. And, of course, there is a special joy in striving and succeeding. So the loss of opportunity for achievement threatens to create an evaluative void. Hence the pressing question: what will be the source of achievement in a world in which work no longer plays that role for many, even the great majority of, people?

Can Games Replace Work as a Source of Achievement?

Perhaps freedom from work will liberate us to devote greater time to spiritual and artistic pursuits, or to family life or democratic governance. Maybe these domains can be sources of achievement that fill the void left by work? But I want to focus on another intriguing proposal that has been advanced by some philosophers, which is that the post-work utopia will be one in which we are primarily occupied in the playing of games.

Games have been defined as the voluntary overcoming of unnecessary difficulties.[12] And if achievement is all about the overcoming of difficulties in realising a goal, then why can't game-playing take the place of jobs as a scene of achievement? This line of argument draws on a conception of game-playing and its value that has been defended by Thomas Hurka. According to Hurka, achievement is the core value of game-playing, and game-playing presents this value in a manner that vividly exemplifies the mindset of modernity, because the value primarily resides in the difficulty of the processes of achieving a goal, not the goal itself; for example, there is no inherent value in putting a ball in a hole (golf).[13] As John Danaher has put it: 'The rise of machines may cause us to retreat to the world of games, but if Hurka is right, this might actually be a good thing because games provide a pure platform for realizing ever higher degrees of achievement.'[14]

Now, it seems to me that there are three problems with this proposal.

The first is a mistake about the nature of achievement. For Hurka, an achievement exists when difficulties are overcome in attaining some goal irrespective of its value. The value of the goal that is attained through a difficult process—for example, a cure for cancer—can magnify the value of the achievement. But for Hurka, achievement exists simply in virtue of the bare fact of overcoming difficulty. I have argued elsewhere that success in a difficult enterprise cannot in itself constitute achievement; instead, achievement requires that difficulties be overcome for a worthwhile goal.[15] Counting blades of grass in well-delineated college quadrangles is not a true achievement, one that gives weight to one's life and merits admiration, however arduous the process. Now, Hurka himself partially admits this point, since he shrinks from the conclusion that there is a genuine achievement when great logistical difficulties are overcome in executing an evil scheme, such as the 9/11 terrorist attack. Success in a difficult activity is not an achievement that makes one's life better if that activity is aimed at very bad ends. But it seems to me that ruling out evil goals is just the more extreme manifestation of the fact that achievement only ever makes one's life go better if it involves success in a difficult enterprise with a worthwhile end, whether that end be the expansion of knowledge, or the maintenance of a friendship, or standing up for justice.

The second mistake is one about the value of playing games. Contrary to Hurka, achievement is not the core value of game-playing; instead, the value of play itself is. On this view, play is a distinct value in itself, apart from the pleasure taken in sports and games.[16] Now, this is not to deny that there may be achievement in playing games, from chess to football. But as with achievement in scientific research (knowledge) or politics (community), the achievement in games is framed by another value (play), which distinguishes them from difficult but pointless activities like counting blades of grass. The independent value of play explains why participation in games can be intrinsically valuable even if that engagement does not exemplify the level of skill needed to count as a real achievement. The rejection of the idea that achievement is the central value of play puts in question the idea that the achievement we find in work can be readily replaced by achievement in game-playing.

Still, someone might object, even if achievement is not the central value of game-playing, games can still be a source of achievement. And isn't that enough for the Hurka-inspired argument to succeed? But now the third mistake in that argument comes into view, which is the idea that the kind of achievement we find in game-playing can readily substitute for the kind of achievement we find in work. In work, unlike in game-playing, we have the opportunity to meet challenges in the course of advancing goals such as health, or justice, or knowledge. The contrast between work and games here partly reflects Johan Huizinga's observation, in his classic study

Homo Ludens, that play is in some sense 'separate from ordinary life', and consequently 'unimportant', in that serious, life-and-death matters do not belong to play.[17] This contrast helps explain why most people would be far more likely to cite their successes in work as among their main achievements in life well before any mention of their golf handicap. Danaher tries to push back against this point by arguing that achievements at work tend to be the preserve of the very few. But this seems to me an elitist distortion. Think, for example, of the philosopher who never makes any original contribution, but who through his teaching inspires a few of his pupils to become philosophers, while inculcating in many others a life-enhancing appreciation of philosophy's value. And a similar case can be made for less abstractly intellectual occupations, from carpentry to taxi-driving.

Work and Natural Reality

Let's add one last twist to this discussion of work and play. Some philosophers—like David Chalmers, as well as John Danaher—go even further and claim that the meaning-conferring games in the techno-utopia will be mainly played in virtual reality. This is because they believe that many or even most of us will spend a lot of our time in virtual worlds, rather than in contact with the physical world, and that we will be no worse off for that.[18] For Chalmers, 'virtual realities have comparable value to nonvirtual realities',[19] hence life in a computer simulation can be just as meaningful as life in a non-virtual world. He thinks this is so irrespective of which of the three leading theories of well-being one favours. Among the ways in which virtual reality might even be superior to non-virtual reality, Chalmers lists the following: they enable us to have experiences that are difficult or impossible in physical reality, for example flying unaided; they are a safe haven if conditions on earth deteriorate; there is no scarcity of resources, for example everyone can own a virtual mansion or planet; and virtual reality is more likely to keep pace with our technologically enhanced minds as they speed up in the future.[20]

Now, Chalmers confronts a number of objections to the claim that life in virtual reality (one in which we may even have 'uploaded' our minds onto a computer) can be just as good as life in non-virtual reality. These objections focus on the absence of such features as relationships, history, and death in virtual worlds, and on the charge that retreat to virtual reality is a kind of escapism. But let me focus here on another of the objections, which I think brings out an important but neglected dimension of the well-being we realise through work. This is the dimension of practical engagement in our original biological human form with the physical world. To this objection, Chalmers says: 'There may be a sense of authenticity in interacting in our original biological form. But it's hard to see why sheer physicality should make the difference between a meaningful life and a meaningless life.'[21]

An immediate objection here is that Chalmers has arbitrarily lowered the bar that his argument must clear. His thesis is that a virtual existence could be just as good as non-virtual existence, which requires much more than that a virtual existence is not meaningless. In this context, it would require showing that the significant lack of contact with the physical world, including our biological bodies, does not in and of itself entail a significant *loss* in value as compared with engagement with virtual reality. Or, if there is such a loss of value, that it can be made up for by the advantages of living a virtual existence.

Now I think it is a strong argument on the other side that among the many goods of work—even allowing that achievement is the central good—is the distinctive form of contact with physical reality that work can afford. Consider this remark in David Wiggins's brief, but characteristically insightful, neo-Aristotelian account of the meaning of work:

> Acts or activities that apply what he [Aristotle] calls a rational principle aim at something worthwhile by drawing upon faculties and dispositions whose exercise gives pleasure (a distinctive, associated pleasure) to the doer and enlarges also—here I reach beyond Aristotle—the doer's understanding of the realities we inhabit. That is to say that the exercise of these faculties or dispositions affords both practical

understanding of those realities and the satisfactions that we attain by learning to wrestle or struggle with them.[22]

The idea here is that work affords a distinct form of understanding that emerges through contact with an independent and potentially recalcitrant physical reality: in work, we engage with that reality in a way that deploys our rational faculties in order to create goods and services that satisfy human interests. This is related to the idea found in Hegel's dialectic of master and slave, which in turn influenced Marx, that work has crucial significance for human self-realisation because it involves the struggle to transform nature into a humanised domain of culture (*Bildung*).[23] By humanising nature in this way we more fully realise our own human potential, which includes enabling us to conceive of ourselves as moral agents possessing equal rights.

At one point, Chalmers remarks that 'Many of us live in cities that are largely human-made, but we still manage to lead meaningful and valuable lives. So artificiality of an environment is no bar to value.'[24] But again this just lowers the bar that his argument needs to clear, which leads to a failure to consider whether some significant *loss* in value is incurred if one were habitually or exclusively living one's life inside an artificial human construct.

Of course, I have only suggested but not shown that such a loss *is* incurred. Still, I want to conclude by proposing that Chalmers misleads us when he asserts that his argument goes through not just on subjectivist theories of well-being like hedonism or desire-fulfilment, but also on an objective list account. On the contrary, it seems to me that it is a subjectivist view of value, according to which value is in some sense the creature of consciousness, that is most likely to make it a matter of indifference whether or not our well-being implicates our biological natures and is rooted in our engagement with a non-virtual reality. Indeed, Chalmers concludes by admitting that his favoured view is that 'all value arises, one way or another, from consciousness.'[25] Still, even if you disagree with me about this, it is clear that one of the gifts of AI technology to philosophy is that it requires us to reflect anew not only on the value of work and play, but on the ultimate nature of human well-being itself.

Acknowledgement

I am grateful to David Edmonds, Michael Ignatieff, and Daron Acemoglu for helpful comments.

Notes

1. Raymond Geuss, *A Philosopher Looks at Work* (Cambridge University Press, 2021), ch. 1. Geuss also identifies three 'less essential' features of work: '(d) a *distinct and almost self-contained activity*, one typically conducted in its own separate space, e.g. a factory, garage, or office; (e) it is *serious*, and hence almost always distinct from what we do for pleasure or fun; and (f) it is *monetised*, i.e. paradigmatically carried out for monetary payment.'

2. Contrast the definition of work adopted by John Danaher: 'Any activity (physical, cognitive, emotional etc.) performed in exchange for an economic reward, or in the ultimate hope of receiving an economic reward.' John Danaher in *Automation and Utopia: Human Flourishing in a World without Work* (Harvard University Press, 2019), 28.

3. C. B. Frey and M. A. Osborne, 'The Future of Employment: How Susceptible Are Jobs to Automation?', *Technological Forecasting and Social Change* 114 (2017): 254–80.

4. James Manyika and Kevin Sneader, 'AI, Automation, and the Future of Work: Ten Things to Solve For', McKinsey Global Institute Executive Briefing, 1 June 2018, <https://www.mckinsey.com/featured-insights/future-of-work/ai-automation-and-the-future-of-work-ten-things-to-solve-for>.

5. For a fine overview, see Roger Crisp, 'Well-Being', *Stanford Encyclopedia of Philosophy*, <https://plato.stanford.edu/entries/well-being/>.

6. Examples of 'objective list theories'—James Griffin, *Value Judgment: Improving Our Ethical Beliefs* (Oxford University Press, 1996); Martha C. Nussbaum, *Women and Human Development: The Capabilities Approach* (Cambridge University Press, 2000), ch. 1; and John Finnis, *Natural Law and Natural Rights*, 2nd edn (Oxford University Press, 2019).

7. Charles Taylor, *Sources of the Self: The Making of Modern Identity* (Cambridge University Press, 1989), 211.

8. Anca Gheaus and Lisa Herzog, 'The Goods of Work (Other than Money!)', *Journal of Social Philosophy* 47 (2016): 70–89.

9. Joshua Cohen, 'Good Jobs', MIT Work of the Future (2020), <https://workofthefuture.mit.edu/research-post/good-jobs/>. See also Jon Elster, 'Is There (or Should There Be) a Right to Work?', in A. Gutmann (ed.), *Democracy and the Welfare State* (Princeton University Press, 1989), 53–78. For a helpful

discussion of the 'bads' of work, including dominating influence, fissuring and precarity, distributive injustice, temporal colonisation, unhappiness and dissatisfaction, see Danaher, *Automation and Utopia*, ch. 3.

10. David Graeber, *Bullshit Jobs: A Theory* (Allen Lane, 2018).
11. John Maynard Keynes, 'Economic Possibilities for Our Grandchildren', in *Essays in Persuasion* (Harcourt Brace, 1932), 358–73.
12. Bernard Suits, *The Grasshopper*, 3rd edn (Broadview Press, 2018).
13. Thomas Hurka, 'Games and the Good', *Proceedings of the Aristotelian Society* supp. vol. 80 (2006): 217–35.
14. Danaher, *Automation and Utopia*, 236.
15. John Tasioulas, 'Games and the Good', *Proceedings of the Aristotelian Society* supp. vol. 80 (2006): 237, 251–60. For an attempt to advance an intermediate position between Hurka's and mine, according to which the value of play sometimes grounds the value of achievement in a game, while in other cases the independently grounded value of achievement in a game provides further grounding for the value of play, see Michael Ridge, 'Games and the Good Life', *Journal of Ethics and Social Philosophy* 19 (2021): 1–26.
16. Tasioulas, 'Games and the Good', 242–51.
17. Johan Huizinga, *Homo Ludens: A Study of the Play Element in Culture* (Beacon Press, 1955).
18. David Chalmers, *Reality+: Virtual Worlds and the Problems of Philosophy* (Allen Lane, 2022).
19. Chalmers, *Reality+*, 328.
20. Chalmers, *Reality+*, 321.
21. Chalmers, *Reality+*, 322. Cf. also the locutions 'don't make the difference between a good life and a bad one', p. 328; 'artificiality of an environment is no bar to value', p. 315.
22. David Wiggins, 'Work, Its Moral Meaning and Import', *Philosophy* 89 (2014): 479.
23. G. W. F. Hegel, *The Phenomenology of Spirit* (Oxford University Press, 1976), 111–19.
24. Chalmers, *Reality+*, 314. Cf. his claim that insistence of physicality may be a 'fetish' if the relevant experiences inside and outside VR are indistinguishable, p. 322.
25. Chalmers, *Reality+*, 329.

Further Reading

J. Danaher, *Automation and Utopia: Human Flourishing in a World without Work* (Harvard University Press, 2019).

PART IV

MANIPULATION, AUTONOMY, AND ALGORITHMS

12

The Silent Meddling of Algorithms

Carina Prunkl

'Your call will be answered in approximately five minutes,' a friendly voice tells you before Beethoven's *Für Elise* comes back on. You are in the telephone queue for Media.INC, the country's biggest tele-communication provider. Your goal is to purchase a broadband package that is both affordable and fulfils your needs. Eventually, you are put through to a young, dynamic lady, called Doris. The two of you get along immediately and you leave the conversation some fifteen minutes later, satisfied and with a new broadband contract. What you don't know: it was no accident that Doris answered your call. There was a reason why you were paired with Doris and not with another agent. An algorithm determined that you and Doris were the 'perfect match'.

Algorithms often operate in the background. In fact, we rarely notice the intermediary function of algorithms in our interactions with institutions or other humans. It is tempting to think that, as long as we are satisfied with the service provided, it doesn't much matter whether an intelligent algorithm was involved in the process. In this chapter, I argue that (i) the use of AI systems in the service sector risks the exploitation of cognitive vulnerabilities of customers and (ii) that even if customers are satisfied with the service, customer satisfaction is a poor indicator for whether the use of an algorithm is warranted.

To kick off the discussion, consider the fictional company *MatchMe*. *MatchMe*'s product is a matchmaking algorithm for call-centres like the one in the opening story: it pairs customer service agents with calling clients on the basis of behavioural compatibility. The technical term for such a matchmaking process is 'Predictive Behavioural Routing' (PBR). The idea behind PBR is to first determine a customer's personality profile and then route their calls to the service agent who is most likely to achieve best outcomes with them. What is considered a good outcome will, of course, vary across contexts: sales made, issues resolved, customers reassured, complaints taken, etc. Central to PBR, however, is the idea that both customer and service agent 'click' during their interactions and that a trusting relationship is established. While *MatchMe* is fictional, there are now an increasing number of such companies on the market. One of them explains in a promotional video on their website: '[…] when you pick up the phone and you feel that on the other side of the line there is someone who understands you—[this] is so important. It is the trust behind this conversation.'[1] Companies like *MatchMe* appear to be incredibly successful on the current market. Clients using the matchmaking software report increased sales and annual revenues in the millions. At first sight, it seems everyone involved is benefiting from the new technology: tech companies are selling their software, client companies are increasing profits, customers are receiving better service.

Yet, there's a catch. The system itself can be unethical.

To understand why, it is helpful to go into more detail about how such a system works. PBR operates with predictive analytics: it draws inferences about the success of future interactions by analysing the success of past interactions. To determine whether a given caller is a good match for a given service agent, the system first requires information on both. In the case of the service agent, acquiring relevant information is relatively straight-forward: the company could look at past agent–customer interactions and perhaps ask its service agents to fill out a questionnaire about personality traits. For calling customers, however, asking them to fill out a questionnaire about their personalities and queries usually isn't an option. If the customer has been in contact before, the system can look for details in past interactions, such as whom she previously talked to, why she called, whether the issue she called for was resolved, and so on. For new customers, the system relies entirely on external databases and live-information. Such information could include, for example, IP address, demographic data, as well as data available on public profiles such as X (formerly Twitter).

Notably, some of this data can be collected in real time: the gender of a caller, for example, can be identified automatically within milliseconds of any given call and can be fed into the system.

With the gathered information on both agents and callers, the system is now in the position to analyse a large number of inter-actions and identify any correlations between the success of a given customer–agent interaction and the particularities of the customer–agent pair. Finally, on the basis of these correlations, the system makes predictions about which caller–agent pair has the highest chance of success and directs incoming calls accordingly. *MatchMe*'s website illustrates the process with the following example: imagine that a particular call-centre agent tends to get better results when talking to older women. The system would then ensure that, where possible, incoming calls from older women are paired with this par-ticular call-centre agent.

Central to the pairing is the idea that customer and agent enter a special relationship when first interacting. If the pairing is success-ful, the customer will perceive this relationship as positive. They will feel at ease and, more importantly, trust the agent to deal with their queries competently. The trust thus established between cus-tomer and agent can be considered a stepping stone for successful customer–agent interactions. The issue here, however, is that trust is not *earned* by the customer service agent. It is generated by the fact that people are more trusting towards some than others. The example above about older women uses age and gender—both pro-tected characteristics—as determinants for trust. The system effect-ively uses cognitive shortcuts to create such trust relationships. That using such shortcuts is morally questionable becomes particularly evident when the objective of the pairing is to maximise revenue, as will be discussed now.

Bad Intentions: Exploiting Cognitive Loopholes

In an ideal world our decisions—including purchasing decisions—are based on careful, rational deliberation. Alas, this is not the world in which we live; instead, our subconscious plays an important role.

A Harvard Business School professor has suggested that up to a mind-boggling 95% of our purchase decision-making could be determined by subconscious processes.[2] While such assertions are difficult to evaluate and are best taken with a pinch of (carefully and rationally purchased) salt, there's undoubtedly some truth in the claim that our decision-making is often influenced by subconscious and irrational factors. In particular, we are frequently biased in our judgements and we are rarely aware of it. Coming back to the call-centre example, this could mean that (statistically speaking) older women are more trusting towards, say, the young, sympathetic female call-centre agent than to the centre's male counterpart. In other words, they are *biased*. Yet, the gender of the call-centre agent should not matter all that much when determining whether or not to trust the person on the other end of the line.[3]

In general, trust, whether it exists for good or bad reasons, is necessary for us to navigate our interactions with institutions or other human beings. Yet it also creates vulnerabilities on the part of the trust-giver. This is particularly problematic if the trust is unwarranted. An obvious example is the context of customer–salesperson relationships, where the objective of the salesperson rarely aligns with the objective of the customer (e.g. maximising profit vs acquiring a product that matches the customer's needs). In these cases, first building trust and then, on the basis of this trust, steering the customer towards spending more money than they had initially planned is a common sales strategy. PBR can provide a short-cut to this standard strategy. Instead of having to create trust in a laborious way, it can identify biases and preferences of customers on the basis of pre-existing data and use them to direct customers to the call-centre agent to whom they are most likely susceptible.

'So what?', the reader might think. Using trust to maximise profit is a practice that is millennia old. Does it really make a difference whether it is an algorithm or a human that facilitates this trust? I believe it does. While common sense, intuition, empathy, and knowledge can go a long way when a salesperson is trying to establish a trusting relationship with a customer, there are both quantitative and qualitative limits to what humans can achieve. On the quantitative side, salespeople don't usually have the same amount of information about a customer as an algorithm, which has access to databases and public profiles that can potentially reveal a great deal about the caller. On the qualitative side, humans are not nearly as proficient as algorithms at analysing patterns and identifying correlations present in large amounts of data. As a result, algorithms are much better at identifying and exploiting biases, or 'cognitive loopholes'. Trained by the right data, algorithms can provide much more immediate, targeted, and accurate predictions about customers.

Good Intentions: Happy Customer—Happy Company?

Let us for a moment put aside the concern that there might be a significant misalignment of objectives between customer and salesperson. Let us instead imagine that the company using the algorithm for its call centres is primarily concerned with customer satisfaction—a reasonable enough assumption to make for many situations. In these cases, trust is needed to ensure that the customer feels heard and understood. The use of a matchmaking algorithm can improve the likelihood of a trusting relationship between caller and call-centre agent being established. But there's a problem: the basis on which such a trust relationship is founded can be morally problematic.

One issue is that the matching of call-centre agents and customers is entirely grounded in data correlations. Given the prevalence of racial, religious, or gender bias in our society, many of these correlations in turn will be based on harmful stereotypes. In the presence of such stereotypes, the algorithm will almost inevitably reproduce and reinforce them. For example, a frequent prejudice against women is that they are incompetent with technology. If this prejudice is mirrored in lower customer satisfaction with female agents, the algorithm will reinforce the prejudice by only forwarding customers with technological queries to male agents. Of course, we could actively counteract the use of certain correlations that result from prejudice and bias, but this is tricky to do, given the complexity of the relationships between features. For example, 'communication style' could be deemed relevant when determining whether or not a customer gets along with a given agent. But what if communication style strongly correlates with gender? In this case, whether we discriminate on the basis of gender or on the basis of communication style—the outcome would be the same. Ideally, we want to make pairing decisions on the basis of relevant *causal* relationships, where the causes themselves are morally permissible (e.g. technical expertise is a permissible cause for pairing; religion is not). But, as shown above, these are not always straightforwardly separable, not to mention detectable.

This is a familiar concern in the literature on algorithms. As already mentioned above, harmful bias and prejudice are so deeply ingrained in our institutions and cultural surroundings that it is almost impossible for us to escape their influence. But there is another angle to the above story: as a customer, I myself might profoundly reject any prejudices associated with race, religion, or gender. Yet I cannot be certain that I have not myself internalised at least some of those prejudices. The matchmaking system that pairs, say, a white, male caller with a white, male customer agent (for the sake of argument we assume that this is a context where such pairing leads to 'optimal' results), makes a decision on the basis of values that the caller himself might very well reject.

Let us assume that a calling customer has some hidden preferences that are based on prejudices to which the customer strongly objects. Had the customer known their preference was based on prejudice, they might object to their own behaviour. Even if the customer ends up being satisfied with the service provided by the call-centre agent, had they been aware of *why* they were paired with a given agent, they would reasonably object to the pairing. Note that this is a distinct yet related concern to the 'bad intentions' issue discussed above. In both cases, the algorithm makes decisions on the basis of subconscious preferences. Yet, in the former case, the morally relevant issue is the exploitation of those preferences, whereas here the concern is that the preferences themselves are objectionable.

There is another dimension to this that relates to company values. Many companies explicitly reject the idea that it is acceptable for customers to act upon blatant prejudice and stereotypes. It would be considered entirely unacceptable in most places for a customer to enter a shop and demand to be served by a white employee. No doubt the customer would be asked to leave, even if this meant a loss of revenue. But using an algorithm that pairs customers on the basis of protected characteristics is doing something similar: it allocates customers to service agents on the basis of stereotypes, bias, and prejudice.[4] Companies therefore face an important decision: should they maximise customer satisfaction at all cost or should they respect basic principles of equality and fairness? The answer, I believe, is obvious.

Concluding Remarks

The above discussion was primarily focused on matchmaking algorithms in a service setting, but it applies more widely. The use of matchmaking algorithms in dating apps, for example, faces similar issues in that users of these apps can experience conflicts between their held beliefs (e.g. 'I can see myself being in an ethnically diverse relationship') and their behaviour (e.g. selecting potential matches on the basis of race) and platform owners need to decide whether they value equality (e.g. each profile is shown equally often on other people's pages) or, say, profit maximisation/customer satisfaction (which might dictate listing the most promising profiles first).

As so often, there is no silver bullet that applies to all contexts. But there are (at least) two relevant considerations that companies, customers, and the public need to take into account when discussing the ethical use of algorithms. First, why are companies deploying algorithms, and are there sound ethical reasons for doing so? Second, why are certain inferences drawn by the algorithms and are they based on stereotypes and historical injustice?

As a customer, you need to be sure that the reason you were put through to Doris neither exploits your cognitive weakness nor reinforces existing patterns of injustice.

Notes

1. <https://www.afiniti.com/>.
2. Gerald Zaltman, *How Customers Think: Essential Insights into the Mind of the Market* (Harvard Business Press, 2003).
3. This is not to say that it is *always* irrational to prefer one demographic group over another in certain contexts and for certain interactions or transactions.
4. Note that this is somewhat simplifying the matter; there are contexts in which it is legitimate to demand being served by someone of a particular social group.

Further Reading

Cathy O'Neil, *Weapons of Math Destruction* (Crown Publishing, 2016).

13
Recommended!

Silvia Milano

In the town where I live, there is a little record shop with a café. The space is too confined to carry an extensive catalogue, but the selection of records is curated by the owner and tends to focus on new releases (often from up-and-coming, local artists). Sometimes a small stage is set up for live music. It is always full of people browsing or sitting in the café listening. Musicians meet there, and it functions as a hub for the local music scene. It's where you go to find new music, or to see what others are listening to.

This kind of setting may sound familiar: it may be a local bookshop, or a knitting-circle, or meeting for wine or motor enthusiasts. It is a setting where people come together, in a more or less structured way, to find and share information about a topic they love. Sharing and giving advice is a pervasive activity that we do implicitly or explicitly across social settings. That's how we learn from others, and find out what we might want to pay attention to.

Good recommendations can extend our horizons and lead us to discover new and interesting things. But other recommendations can be less successful. For instance, if the coffee shop only played the most popular songs, customers might miss variety and character. Listeners might value being challenged—as opposed to having their current habits reinforced.

The local café is a hub both for recommendations and for facilitating recommendations. What if that function were automated? So called 'recommender systems' automate the activity of advice-giving, and have enjoyed enormous popularity since the advent of the internet. Without a system to structure the ~16,000 videos uploaded on TikTok every minute, for example, it would be impossible to find anything worth watching. Human curation would be impractical due to the sheer volume of content—so recommender systems have become essential instruments to help reduce information overload. But even in contexts where the catalogue of options is smaller, automated systems offer new ways to personalise recommendations to individual users—for example Spotify's 'Made For You' mixes—at a scale that would not be feasible for human curators.

The modern approach to automated recommendations began in 2006. Netflix, then still a video rental service, offered a cash prize of $1 million to the engineering team that could improve the accuracy of its recommendation algorithm by at least 10%. The competition ignited research interest in the area, setting up a pattern: researchers (in and outside industry) would look at existing datasets and experiment with novel approaches to extracting information about user preferences and the items (e.g. movies, books, news reports, social media posts, etc. depending on the application) with which they interacted.

Recommendation and Inference

In abstract terms, the task assigned to a recommender system can be described as: 'find good items' or 'predict an item's relevance to a user',[1] from among a set of available options. The system tries to learn a user's preferences based on past data. In other words, it performs an *inference task*. The success of the resulting recommendations may depend on several factors, including the quality and quantity of the available data, the method of analysis, and the context in which the recommendation is given. For example, the quality of music recommendations available on a streaming platform for a user might be influenced by how much the user has interacted with the platform in the past (the greater the interaction the more data points the algorithm has to tailor personalised recommendations), and by how similar their listening habits are to other platform users.

The fact that recommender systems are based on inference suggests a parallel to scientific inference. Much as weather forecasting is based on historical data, so recommender systems base their predictions on the data that is available to them. In the case of recommendation, data about users, items, and interactions between them, as well as what is known about context, carries information that is relevant to predicting a user's preferences or what recommendations they're likely to respond to. The use of machine learning to extract patterns and analyse hidden or surprising correlations encourages a voracious approach to data collection. All data is seen as potentially useful: for instance, a user's mouse movements on the page, browser history, or time of day are analysed for correlations with purchasing behaviour (from similar users), in the hope that it might reveal which options the user is most likely to react to.

The parallel between recommendation and scientific inference highlights two important aspects of the way in which 'recommender systems' are approached, which we'll call 'the conventional view'. On the conventional view, the main objective of recommendation is accuracy; that is, correctly predicting user preferences. Accuracy is also viewed as neutral with respect to other normative or practical values, such as good taste or moral praiseworthiness. We can contrast this with the coffee shop, where the owners care about and learn from their customers' tastes, but might also play music just because they like it.

Let's consider an example to illustrate the conventional view. Suppose we want to build a recommender system to suggest podcasts on a streaming platform. The data to train the recommender system might include the information on the users' profiles and listening history, including time and location, podcasts liked and hosts or topics that users follow, and any other linked social media accounts or friend networks. All this data might be used to predict the preferences of users and show them podcasts that they might like. From the point of view of a user, a recommendation is more valuable if it brings to light podcasts that they want to listen to.

That accuracy is a fundamental value for recommendations is reflected in the use of accuracy metrics, which are by far the most common way of evaluating a recommender system.[2] The value of accuracy is seen as independent of any judgement about the quality of the tastes of users, or whether the items they prefer are profitable for the platform. While ethical considerations may mean that sometimes we shouldn't recommend what a user would actually prefer (e.g. there may be reasons to moderate content and not promote podcasts that contain politically inflammatory content, even if some users would like to listen to it), these considerations are usually viewed as constraints on accuracy. In practice, this means that they are taken into account when *moderating* the items that are recommended, after predictions about a user's preferences have been made.[3]

The assumption of neutrality is often also applied to the data on which the recommendations are based. When recommender systems are designed, there may be some data that can't be collected or used, due for instance to privacy concerns or data protection laws. However, absent such limitations, it is commonly thought that all data is potentially useful and informative. In theory, the more data that is fed into a model, the more accurate the prediction.

Where Things Go Wrong

Recommender systems are known to raise several challenges, including risks to the privacy, autonomy, and welfare of their users, as well as problematic social effects. As we'll see, the focus on accuracy as the goal of recommendation is a source of, or exacerbates, many of these issues.

First, recommender systems, especially those based on 'collaborative filtering' algorithms, have a tendency to give rise to problematic feedback loops. Collaborative filtering is a technique that leverages the information contained in how other users rated or interacted with items recommended, to make a prediction about how a new user will respond to a recommendation.[4] The idea is that if many people interacted positively with an item (e.g. liked a song), then the item is more likely to be of better quality or more relevant than other items that were not rated as much.

The same reasoning can be applied within smaller groups. For example, some listeners may be more interested in history podcasts, while others prefer true crime. These preferences can be independent of the intrinsic quality of the items, but are quite predictable once we know what group a user is more likely to fall into, which is something that recommender systems can be good at detecting. The record shop in my town, for instance, seems to be a hub for alternative indie and electronic music, and this is reflected in the curated record selection. Online, similar communities can grow around blogs or other virtual spaces. Recommender systems can recreate a version of this by analysing user behaviour, and recommending content that was liked by similar users, interpreting each positive interaction as a 'vote' of preference. As the inferred judgements are based on many individual 'votes', they can be more accurate and robust than other forms of inference. It is akin to tapping into the 'wisdom of the crowd', but where the goal of prediction is not a general statement (e.g. whether a podcast is good in absolute terms), but rather an individualised prediction (whether a user with a specific profile will like a podcast).

One of the advantages of this method is that it makes it possible to leverage the distributed knowledge of users, without imposing pre-set judgements about the relevance of different options, since the developers of the recommender system may not be knowledgeable about the items to be recommended. For example, we could use this method to learn what podcasts similar users have listened to, and use that information to formulate a personalised recommendation, without interrogating why they liked it. However, a drawback is that it is still prone to generating recommendations that reinforce the popularity of already popular items, at the expense of variety. This degrades the quality of recommendations over time, since more obscure items are less likely to be recommended, preventing users who might have liked them from discovering them. Moreover, the presence of feedback loops risks placing users in filter bubbles or echo chambers, where their beliefs and tastes are insufficiently challenged or simply reinforced.[5] This is a particular problem in contexts where recommender systems influence what news or political views their users are exposed to.

Mitigating the issue of feedback loops in recommender systems requires considering other values besides accuracy of prediction, and, importantly, recognising that recommendations are not neutral with respect to other values. If, for example, it is of intrinsic value in a democratic society that citizens engage with different views in the political arena, recommender systems for news media need to carefully balance accuracy with diversity of exposure.[6] This example illustrates how recommendations can have social value beyond the individual utility to users.

An additional concern raised by recommender systems is that they may encroach on the autonomy of their users: they might manipulate and influence us, and in ways that may be difficult to detect. Getting together with a group of music-loving friends or sharing tracks at the local record shop are seen as positive ways of developing and deepening tastes in music. But it is more difficult to trace the influences in one's development when this happens online and is mediated by an algorithmic system. This raises the worry that online platforms could use their recommender systems to manipulate users for their own gain, a worry that is especially pressing given the prominence of online advertising and scandals involving political communication on social media.[7]

An interesting aspect of the issue of manipulation is that, by their nature, manipulative recommendations may increase accuracy. If the preferences of a user can be manipulated, this would also make them easier to predict. For example, if we can make our recommender system for podcasts sufficiently persuasive, this may in turn increase the accuracy of the recommendations it makes. However, the value of this accuracy improvement would be debatable, at best.

Can Recommendations Be *Too Good?*

Imagine being a regular at my local café/record shop, and each day the owner recommends things eerily well. Perhaps you're depressed following an argument, or, unbeknownst to the owner, are happy about a promotion at work, yet somehow the café always plays the perfect record for your current mood. How would you feel about this? You might start to wonder how the owner knows so much about you.

From the point of view of users, the receivers of a recommendation, recommendations can seem problematic both when they are intuitively off (why am I being recommended THIS!), but also, more puzzlingly, when they're right in suspect ways. For example, users have reported feelings of uneasiness after receiving recommendations for things that they were interested in, but felt that the inference made by the algorithm was too personal, or based on information that should have been private.

Some women have reported receiving recommendations for baby products before even realising that they were pregnant, or after suffering a miscarriage.[8] This raises obvious risks, for example if others learn sensitive information from seeing the personalised recommendations for a user. Some women may be wary of others accidentally noticing the targeted ads for pregnancy products popping up in their browser at work, as this might reveal sensitive information that they'd prefer to keep private.

The conventional view, centred on the accuracy of recommendations, does not capture what is at issue in these cases. While it is relatively easy to explain what is wrong with inaccurate recommendations, the fact that some recommendations may be problematic despite, or even precisely because, they are accurate, should give us pause. One possible explanation is that, as mentioned before, such accuracy may indicate that the recommendations are manipulative. However, some of the problematic cases do not involve manipulation, and therefore present a subtler challenge. These are recommendations that appear to pick up on something that is relevant to the user, but come almost as unwelcome advice and an intrusion on privacy.

A better explanation should draw on other values that recommendations (and advice-giving as a general practice) have from the point of view of those receiving them. Accuracy is without doubt an important value, but supporting users' autonomous choices is arguably as essential, since ultimately the function of good recommendations is to help the people receiving the recommendations better navigate complex information environments—and not to automate away their freedom of choice.

Values in Recommendations

The focus on the accuracy and neutrality of predictions reflects a common narrative about how so called 'Big Data' has revolutionised science. In the words of one writer, the deluge of data available to scientists 'has made theory obsolete',[9] profoundly changing the nature of scientific enquiry. Other authors have pushed back on this triumphalist narrative,[10] pointing out the limitations of purely data-centric approaches, and the need for theory to underpin both the data collection process, and the quality of scientific inferences.[11]

Centring the value of recommendations in supporting the agency of their users should lead us to revise the conventional view of the recommendation problem. Writing about the role of values in science, the philosopher Helen Longino forcefully argued that the excessive focus on accuracy is at odds with the provision of relevant and useful knowledge. According to Longino, scientific theories position different parties as more central to the questions and methods of enquiry, influencing the distribution of power. For example, research hypotheses and experimental treatments in the engineering sciences traditionally assumed the healthy adult male as the reference. Translated into safety testing, this resulted in products that were unsafe for women and children. The assumption of a canonical perspective also results in privilege for some groups when accuracy is the focus of recommendation, systematically disadvantaging minority groups whose interests and preferences are served less well.[12]

Longino's observations help illuminate a source of problems with the conventional view of recommendation. Recommender systems function by centralising the data and inferences about their users, but, along with many algorithmic systems, they are opaque, and take away power from their users. The recipients of recommendations have little ability to understand the reasons behind them, nor how and what data was used to reach them. Recommender systems depend on the availability of data, favouring the interests of the platforms that make the decisions to develop them. Undeniably, the users of recommender systems can derive a benefit from the recommendations, which are often sold as a service. However, the choice to build a recommender system, and the decision to give it a specific interpretation of what constitutes an accurate recommendation, is solely within the remit of the platform—meaning that its users, the objects of the system's predictions, have no control over the enquiry. This state of affairs is especially problematic, since it interferes with the ideal goal of recommendations to support the free choices of its recipients.

*

We can now see that a problematic aspect of some accurate recommendations is that they provide suggestions without insight into how or why they were produced. Their accuracy may make you more predictable to the online platforms that you use, but does not give you power to scrutinise when or why you are profiled. Perhaps the ready availability of commercial recommendations makes them more enticing for users; in other areas of our lives fewer technological instruments are available. When the availability of accurate predictions is selective, the question arises as to why attention is directed here.

Our response should be to focus on redistributing power in the practice of developing recommender systems. The availability of recommender systems for certain domains channels attention to those domains, at the expense of others. Accuracy should be an instrumental goal for recommendations, only on condition that the purpose of the advice is sound: the reason why accurate recommendations can be good, is that they can empower receivers to identify relevant options and act on their own preferences. However, as we have seen, accuracy is not empowering when it leads to negative feedback loops, when recommendations are manipulative, or when the accurate predictions are intrusive and unwanted.

Diffusion of power, in all those cases, is difficult because the centralised nature of recommender systems means that they are too complex for individuals to understand how they work. A first step would involve making the systems more transparent, especially with respect to how the performance of the recommender system is evaluated. But on a deeper level, recognising that the focus on accuracy raises all the issues that we have seen, we will need different ways to evaluate recommendations—that is, we will need other values. This perspective might seem challenging: how will we select values for recommender systems, without engaging in wishful thinking?[13] And how will we avoid the capture of these values by tech companies or vested interests? These are crucial and difficult questions, which I do not attempt to answer here. But just as important as the values is the process by which we choose them. The best way to approach this problem may be through a democratic process, providing space to publicly vet what recommender systems are built, and with what values.

In thinking of recommender systems more like a public infrastructure than a private service, we might reach the conclusion that some recommender systems have a public value and support our individual and collective interests in their domains, allowing us to tap into the distributed knowledge of others to make better choices. In other domains, however, we might reach the conclusion that accurate recommendations do not serve our interests and are at best distracting. In those cases, we have reasons to refrain from developing them.

Acknowledgement

I would like to thank Adrian Currie and Milo Phillips-Brown, in addition to the editor of this collection, for their generous comments on earlier versions of this chapter.

Notes

1. Dietmar Jannach and Gediminas Adomavicius, 'Recommendations with a Purpose', in *Proceedings of the 10th ACM Conference on Recommender Systems—RecSys '16* (presented at the the 10th ACM Conference, Boston: ACM Press, 2016), 7–10, <https://doi.org/10.1145/2959100.2959186>.
2. Nava Tintarev and Judith Masthoff, 'Explaining Recommendations: Design and Evaluation', in Francesco Ricci, Lior Rokach, and Bracha Shapira (eds), *Recommender Systems Handbook* (Springer US, 2015), 353–82, <https://doi. org/10.1007/978-1-4899-7637-6_10>; Francesco Ricci, David Massimo, and Antonella De Angeli, 'Challenges for Recommender Systems Evaluation', in *CHItaly 2021: 14th Biannual Conference of the Italian SIGCHI Chapter*, CHItaly '21 (Association for Computing Machinery, 2021), 1–5, <https://doi.org/ 10.1145/3464385.3464733>.

3. Paddy Leerssen, 'An End to Shadow Banning? Transparency Rights in the Digital Services Act Between Content Moderation and Curation', *Computer Law & Security Review* 48 (2023): 105790, <https://doi.org/10.1016/j.clsr.2023.105790>.

4. Yehuda Koren, Steffen Rendle, and Robert Bell, 'Advances in Collaborative Filtering', in Francesco Ricci, Lior Rokach, and Bracha Shapira (eds), *Recommender Systems Handbook* (Springer US, 2022), 91–142, <https://doi.org/10.1007/978-1-0716-2197-4_3>.

5. C. Thi Nguyen, 'Echo Chambers and Epistemic Bubbles', *Episteme* 17/2 (2020): 141–61, <https://doi.org/10.1017/epi.2018.32>.

6. Elliot Jones, Catherine Miller, and Silvia Milano, *Inform, Educate, Entertain… and Recommend? Exploring the Use and Ethics of Recommendation Systems in Public Service Media* (Ada Lovelace Institute, 24 November 2022), <https://www.adalovelaceinstitute.org/report/inform-educate-entertain-recommend/> (accessed 7 February 2023).

7. 'Social Media Manipulation by Political Actors an Industrial Scale Problem—Oxford Report | University of Oxford', <https://www.ox.ac.uk/news/2021-01-13-social-media-manipulation-political-actors-industrial-scale-problem-oxford-report> (accessed 22 July 2021).

8. Rae Nudson, 'When Targeted Ads Feel a Little Too Targeted', *Vox* (2020), <https://www.vox.com/the-goods/2020/4/9/21204425/targeted-ads-fertility-eating-disorder-coronavirus> (accessed 28 May 2023).

9. Chris Anderson, 'The End of Theory: The Data Deluge Makes the Scientific Method Obsolete', *Wired*, <https://www.wired.com/2008/06/pb-theory/> (accessed 26 May 2023).

10. Sabina Leonelli and Niccolò Tempini (eds), *Data Journeys in the Sciences* (Springer International Publishing, 2020), <https://doi.org/10.1007/978-3-030-37177-7>.

11. Helen E. Longino, *Science as Social Knowledge: Values and Objectivity in Scientific Inquiry* (Princeton University Press, 1990); Elizabeth Anderson, 'Knowledge, Human Interests, and Objectivity in Feminist Epistemology', *Philosophical Topics* 23/2 (1995): 27–58.

12. Michael D. Ekstrand and others, 'Fairness in Recommender Systems', in Ricci, Rokach, and Shapira, *Recommender Systems Handbook* (2022), 679–707, <https://doi.org/10.1007/978-1-0716-2197-4_18>; Himan Abdollahpouri, Masoud Mansoury, Robin Burke, and Bamshad Mobasher, *The Connection Between Popularity Bias, Calibration, and Fairness in Recommendation*. In Proceedings of the 14th ACM Conference on Recommender Systems (RecSys '20), 2020. Association for Computing Machinery, New York, NY, USA, 726–731. https://doi.org/10.1145/3383313.3418487.

13. Anderson, Elizabeth. 'Knowledge, Human Interests, and Objectivity in Feminist Epistemology'. *Philosophical Topics* 23, no. 2 (1995): 27–58. http://www.jstor.org/stable/43154207.

14

Do AI Systems Allow Online Advertisers to Control Others?

Gabriel De Marco and Thomas Douglas

Many are concerned about the use of AI systems in search engines, social media, online advertising, and recommendation systems. One worry is that these systems are contributing to bad outcomes, like low vaccination rates or the election of populist and divisive politicians. Another worry is about the relationship between the companies employing the systems and the individuals affected by them. Some are apprehensive, for instance, about companies deploying AI systems that manipulate, deceive, or control the users of their platforms. Here are some representative excerpts:

> As long as [data scientists] have access to our data we will continue to be their puppets. The only way to take back control of our autonomy, our ability to self-govern, is to reclaim our privacy.[1]

> We're being tracked and measured constantly, and receiving engineered feedback all the time. We're being hypnotized little by little by technicians we can't see, for purposes we don't know. We're all lab animals now.[2]

> Now people have become targets for remote control, as surveillance capitalists discovered that the most predictive data come from intervening in behaviour to tune, herd and modify action in the direction of commercial objectives.[3]

In this chapter, we focus on the issue of control, specifically in relation to AI-based targeting of *advertisements*. We consider how AI-systems may affect the degree to which online advertisers have control over those who view the advertisements. Our ultimate interest is in interpersonal control—control of one person's thought or behaviour by another person. However, since most existing work on control has examined instead the way in which one individual controls her own behaviour, we start with that.

On Control

Rory McIlroy, one of the world's greatest golfers, kneels down to touch the green. He passes his hand through the grass, feeling its length and thickness. He squints and carefully observes the slope of the green between his ball and the hole. He selects a putter and asks his caddy to bring it to him. Having grabbed the putter, he stands, sets his legs, checks his grip, pulls back slightly, and strikes the ball. The ball rolls along, curving with the slope, and just before it reaches a full stop, drops into the hole. Had the ball initially been a bit further from or closer to the hole, or the slope been slightly different, Rory would have adjusted accordingly; hitting the ball harder or softer, or at a different angle to account for the slope. Rory knows what he is doing, and he has great control over the ball, and over whether it goes into the hole.

What is involved in this control? One of the things we associate with control is success. A person who has great control over what they are doing has a higher likelihood of success than someone who has little control over it. Rory McIlroy has a higher likelihood of success over a given putt than either of us does. Further, this higher likelihood of success applies in a large variety of circumstances; regardless of where the ball is on the green, McIlroy would have a higher chance of sinking it than either of us.

Having a high degree of control in this example requires having a variety of abilities. Consider some of the most important.

First, McIlroy can recognize features of his circumstances that are relevant to achieving his goal: putting the ball into the hole. Not only is he a good judge of the distance between his ball and the hole, he is a good judge of the slope and what effect it will have on the ball as it moves along the green, and he knows how features of the grass will affect the ball as it travels, given its length, thickness, and the firmness of the ground.

Second, he has the ability to form a plan whose execution is likely to achieve this goal. He picks the putter that will be best in these circumstances, and determines how hard he has to hit the ball, and in which direction. Had some of these features been different—say, had the slope been steeper—his plan would have been different as well: he would have aimed in a different direction.

Third, he has the ability to implement various parts of this plan. For example, he can swing the putter in just the right way to transfer the right amount of force, in the right direction. This ability is critical for control. Someone who is very good at recognizing the relevant features of their circumstances, and can plan accordingly, may not have much control if they are not good at implementing their plans.

Finally, as he is implementing his plan, McIlroy can monitor his performance and correct it if need be. When he sets his feet, he double-checks to make sure they are in the right position, and he can adjust his stance if they are not. If there is a gust of wind, he can pause and wait for it to pass.

Online Advertising

These abilities are a part of what it is to control something. In the example above, they were relevant to McIlroy's control over an object: the golf ball. But people can also have, and exert, control over each other. And one way they might do so is via the use of advertisements or other marketing techniques.

Consider, first, a basic form of online advertisement. Adidas offers a large variety of products designed for many different activities. Suppose, then, that Adidas offers different ads on different parts of the website of cable sports channel ESPN. It may, for instance, have an ad for basketball shoes featuring James Harden (shooting guard for the Philadelphia 76ers) and a snappy, inspirational quote on the ESPN basketball pages, while showing ads for football boots featuring Lionel Messi on the football pages.

In implementing this strategy, Adidas displays some of the abilities mentioned above, if only to a minor extent. In order to increase sales of Adidas products, it tracks features of ESPN readers—for example, that they are likely to recognize Harden and Messi, and that they are interested in sport—which may be relevant to success in increasing sales. It forms a plan on the basis of this recognition; it includes the most popular basketball or football player, and the quote that people rated as best in focus groups. And it implements that plan; it contacts ESPN and sets the process in motion. Had things been somewhat different—had a different Adidas-sponsored athlete been more popular, or had a different quote been rated more highly—the ad may have featured that athlete, or that quote, instead.

This strategy also involves a bit of personalization, in that which ad a person sees depends on whether they visited a page for basketball or for football. In this respect, it is an improvement over the simpler strategy of just showing the Harden ad for basketball shoes on all ESPN pages, since it implements a different plan— that is, shows a different ad—depending on the feature of the individual—that is, which site they visited, and which sport they are interested in.

Presumably this personalization or targeting makes the advertising campaign more effective. Yet, the campaign is still fairly indiscriminate. There can be vast differences among those who visit the same page, and these differences can affect the likelihood of success with regards to any particular individual. Some may follow professional basketball but have no interest in playing it, some might be diehard Nike fans, some might not be in the market for new shoes, etc.

AI-assisted Advertising

Adidas' strategy, described in the previous section, is limited in two ways. First, the features that Adidas picks up on are only somewhat related to what Adidas really wants to achieve—selling more of its products; it misses certain features that are relevant to this, for example, whether the basketball fan actually plays basketball herself, and whether she needs a new pair of shoes. Second, in our example,

Adidas has only a limited range of ads that it can show to different people. AI systems might, to some extent, be able to overcome both of these limitations, and it is this that has caused some to worry that AI-targeted ads may amount to a form of control.

With respect to the first limitation, AI, and more generally, automated systems, could help in various ways. One reason that Adidas is limited in the features it can track concerns the number of data-points that can feasibly be gathered. When a campaign is intended for a vast number of individuals, humans cannot reasonably gather and analyse much data about the different individuals, or at least not in a cost-effective way. Automated systems can.

Take Facebook, for instance. Facebook tracks all of the data we explicitly give when creating a profile, as well as what we like (or react differently to), what we share, who our friends are, and what videos we watch while on the site. It has even tracked our mouse activity,[4] and things we write and decide to delete before posting.[5] Or consider data brokers, which gather data about individuals across a variety of platforms or interactions. These brokers often synthesize data, matching data about individuals from different sets to create a more comprehensive profile of individuals; profiles which they can then sell.

AI systems can analyse all of this data to find correlations, and improve predictive power. Here is a simple example. Suppose we have some further information about two individuals who visited the ESPN basketball page. One of these individuals, call her Rachel, has recently been browsing women's basketball shoes on different sites. Another, call him Pedro, has recently purchased a pair of basketball shoes. If we had this information and, let us assume, we had different sorts of ads available, then we might present an ad for women's basketball shoes to Rachel, and an ad for some other related accessory (head-bands or sweat-shirts) to Pedro.

This example makes use of features that are intuitive to us. Yet there is a vast array of other features, some of which may not be as intuitive, which can affect the likelihood of a person's purchasing an Adidas product. An AI system can analyse data about individuals, including data about who purchases items after exposure to an ad, to find more of these features. The more of such features it can find, the more precise its profile for a particular individual, the better it can get at picking the right sort of ad to show the individual. In this way, AI could help improve the success rates of the ad campaign.

The second limitation mentioned above was that there is a limit in terms of 'plans' available. In our example above, Adidas just had one ad per sport. Limited in this way, it cannot do much with its information regardless of how much it has. To return to the golfing case, Adidas would then be in the position of a physicist who can take accurate measurements of the ball's distance from the hole, the angle of the slope, the firmness and length of the grass, etc., but who is a poor putter because she cannot adjust her movements precisely in response to these features.

The number of different plans—that is, ads—available may differ with the resources the advertiser has available. A large company like Adidas can create a large number of ads, even though they are expensive to produce; a small business or local store might not be able to make as many. Can AI help with producing new plans, or ads, say, by creating content? Yes. Companies such as Jasper, Mentum, and Rocket Content offer access to AI that will write content for ads, blog posts, or articles, or create images that can be used in messaging. Or consider a service offered by Meta, the owner of Facebook, Instagram, and WhatsApp, called Dynamic Creative Ads.[6] This service allows you to provide them with multiple components of ads—text, images, videos, audio, etc.—and it will use them to create 'variations for each person who views your ad.'[7] A further feature allows the system to automatically modify some of the components you have provided, or even add new content, like songs. They do this in a way intended to maximize performance.

Control and Success

The use of artificial intelligence can help companies surmount the limitations of the more traditional strategies we mentioned above. And it is these features of AI-assisted advertising that are typically referred to when others raise the concern that advertisers are controlling people. However, although we agree that AI-assisted advertising can help to enhance some of the abilities relevant to control, here we offer a reason to hesitate before accepting that these improvements are enough for control.

Recall that control is associated with success. If someone has greater control over what they are doing than another person—for example, McIlroy's putting compared to ours—then they will tend to have a greater likelihood of success. If someone has a low likelihood of success, then they do not have control; or, at least, this would cast doubt on the claim that they do. The four abilities we mentioned at least partly explain why someone is likely to succeed. But one can have all of these abilities to some degree while still having a low likelihood of success and, so, we think, while still lacking control.

Consider the first strategy we considered, where Adidas presents relatively indiscriminate ads—one for basketball shoes, one for football shoes, etc.—on the web pages for different sports. In neither of these cases does Adidas have control over the visitors to the website, or over what they do. This is partly because, although advertising might increase the likelihood of success, viewers of the ad are still unlikely to buy Adidas products as a result of it. The ads either have a very small effect on viewers, or affect only a few of them.

Artificial intelligence can help to enhance the relevant abilities and increase the likelihood of success. But if the increase is not great, and the likelihood of success remains relatively low, then again, we do not see how this increase in likelihood can grant advertisers control over individuals. One relevant question, then, is how successful such AI-assisted marketing is. There are varying estimates, but current research suggests that the answer is 'not very'.

Consider, for example, one of the major studies done to test this.[8] This study involved using personal data from Facebook to predict whether an individual is more introverted or extroverted, and then showed them an ad designed for introverts or extroverts, respectively. Though the targeted ads were somewhat more effective than non-targeted ads—subjects were 1.54 times as likely to make a purchase if they saw a targeted ad—the number of people who made a purchase after viewing the targeted ad was still miniscule—390 out of over 3 million people exposed to it.[9]

So, although AI systems can increase the success rate of ads, and they do so via the improvement of some of the capacities relevant to control, there is still reason to think that such methods do not confer control. Ultimately, there may just be a fairly low limit on the likelihood of advertising success; it may be that, no matter how personalized the content of an ad, it will rarely succeed. If this is right, it is hard to see how such ads can grant advertisers control over individuals.

Conclusion

To conclude, while we agree that modern methods of data gathering and analysis raise serious issues unrelated to control, and we also think there is something problematic about *attempts* to control individuals by using these technologies to target ads, we doubt that these technologies currently enable advertisers to control ad-viewers. We should acknowledge, though, that we have focused on one type of advertising—commercial advertising—and on individual instances of it. Perhaps receiving a variety of more personalized ads for some products over a long enough period would have a higher success rate. And perhaps some other forms of advertising—for example, political advertising—are more effective, for example, because of the way they interact with other influences, such as news stories. The effects of political advertising, however, are less clear than for commercial ads, partly because there are so many factors in play that it is more difficult to figure out what affects voting behaviour, and by how much.

There is also the question of whether AI systems allow online actors, such as social media platforms, to control users not via advertising, but via the techniques they employ to keep users engaging with the platform itself. They may do this, for example, by serving up targeted content that users are likely to find especially salient, controversial, or interesting.

This question is not addressed by our arguments above and it is not clear whether concerns about control are more or less plausible in relation to these techniques than in relation to advertising. On the one hand, the goal of keeping a user engaged is much more open-ended, and can be achieved in many more ways—continued scrolling, reacting to content, logging in again—than the goal of a company like Adidas: selling more Adidas products. For such an open-ended goal, a platform like Facebook may have many more plans available, insofar as there is much more content, and many more features, at their disposal to help them achieve this goal. On the other hand, the efficacy may still be fairly low. Further, since this will very often involve finding content that the user is already interested in, this may be less of a case of Facebook imposing some specific goal on an individual. The individual's interests arguably play a larger role in determining what content they see.

Compare this to the Adidas strategy. Although that strategy can personalize to some extent, and which ad a person sees may be partly determined by their interests, it still requires content that is related to Adidas's very specific goal of selling more products. This difference may suggest either that Facebook's strategy does not involve more control over individuals, or, if it does, such control may be less concerning.

In any case, we have raised doubts about the claim that AI-assisted advertising already gives companies like Adidas control over ad-viewers. Although AI-assistance can enhance the abilities that such companies have—for example, by gathering and analysing massive amounts of data, by developing a wider range of potential ads, and by finding better matches between viewers and particular ads—such improvements cannot confer control if the ads themselves are largely ineffective.

Notes

1. Carissa Véliz, *Privacy is Power* (Penguin Random House, 2020), 71.
2. Jaron Lanier, *Ten Arguments for Deleting Your Social Media Accounts Right Now* (Penguin Random House, 2018), 5.
3. Shoshana Zuboff, 'You Are Now Remotely Controlled', *New York Times*, 24 January 2020.
4. See e.g. Michael Grothaus, 'Forget the New iPhones: Apple's Best Product is Now Privacy', *Fast Company*, 19 September 2018.
5. Casey Johnston, 'Facebook is Tracking your "Self-censorship"', *WIRED*, 17 December 2013.
6. For an overview of this offering, see <https://www.facebook.com/business/help/170372403538781?id=244556379685063>.
7. Ibid.

8. S. C. Matz, M. Kosinski, G. Nave, and D. J. Stilwell, 'Psychological Targeting as an Effective Approach to Digital Mass Persuasion', *PNAS* 114/48 (2017): 12714–19.
9. For a more in-depth discussion of this study, and micro-targeting in general, see Y. Benkler, R. Faris, and H. Roberts, *Network Propaganda* (Oxford University Press, 2018), 275–9.

Further Reading

Thomas Christiano, 'Algorithms, Manipulation, and Democracy', *Canadian Journal of Philosophy* 52/1 (2022): 109–24.

15

Should You Let AI Tell You Who You Are and What You Should Do?

Muriel Leuenberger

Your phone and its apps know a lot about you. Who you are talking to and spending time with, where you go, what music, games, and movies you like, how you look, which news articles you read, who you find attractive, what you buy with your credit card, and how many steps you take. Personal information about individual preferences, characteristics, and actions has turned digital. Nearly everything you might want to know about a person is available or can be inferred from stored 1s and 0s. This information is already being exploited to sell us products, services, or politicians. Online traces allow companies like Google or Facebook to infer your political opinions, consumer preferences, whether you are a thrill-seeker, a pet lover, or a small employer, how probable it is that you will soon become a parent, or even whether you are likely to suffer from depression or insomnia.

With the use of artificial intelligence and the further digitalization of human lives, it is no longer unthinkable that AI might come to know you better than you know yourself. The personal user profiles AI systems generate could become more accurate in describing their values, interests, character traits, biases, or mental disorders than the user themselves. Already, technology can provide personal information that individuals have not known about themselves. Such elaborate personal profiles might enable AI to tell you what you really want—not just which movies or music fit your taste—but who you want to vote for, which job would suit you, or which potential partner you should date. Yuval Harari[1] exaggerates but makes a similar point when he claims that it will become rational and natural to pick the partners, friends, jobs, parties, and homes suggested by AI. AI will be able to combine the vast personal information about you with general information about psychology, relationships, employment, politics, and geography, and it will be better at simulating possible scenarios regarding those choices.

Knowing yourself and improving your life choices is advantageous in many ways. Self-knowledge is instrumentally useful since it saves a lot of time and stress by helping you to make plans and decisions that are more likely to enhance your well-being. It is also good because it makes you a more reliable cooperator. What's more, you will cost the health system less if you know more about your body and mind. Some philosophers even argue that you owe it to yourself to know yourself. According to Kant, for instance, this duty to oneself grounds all moral duties. Self-knowledge mitigates the tendency to deceive yourself into believing that what you want is morally right. Even if AI did not increase self-knowledge but helped you make better personal decisions without fully understanding them, your well-being would likely increase. Society at large would probably also profit, because you would, for instance, vote for a political party that actually represented your interests instead of the one with the bigger advertisement budget.

So it might seem that an AI that lets you know who you are and what you should do would be great, not just in extreme cases, à la Harari, but more prosaically for common recommendation systems and digital profiling.

I want to raise two reasons why it is not.

Trust

How do you know whether you can trust an AI system? How can you be sure whether it really knows you and makes good recommendations for you? Imagine a friend telling you that you should go on a date with his cousin Alex because the two of you would be a perfect match. When deciding whether to meet Alex you reflect on how trustworthy your friend is. You may consider your friend's reliability (is he currently drunk and not thinking clearly?), competence (how well does he know you and Alex, how good is he at making judgements about romantic compatibility?), and intentions (does he want you to be happy, trick you, or ditch his boring cousin for an evening?). To see whether you should follow your friend's advice you might gently interrogate him: why does he think you would like Alex, what does he think you two have in common?

This is complicated enough. But judgements of trust in AI are more complicated still. It is hard to understand what an AI really knows about you and how trustworthy its information is. Many AI systems have turned out to be biased—they have, for instance, reproduced racial and sexist biases from their training data—so we would do well not to trust them blindly. Typically, we can't ask an AI for an explanation of its recommendation, and it is hard to assess its reliability, competence, and the developer's intentions. The algorithms behind the predictions, characterizations, and decisions of AI are usually company property and not accessible by the user. And even if this information were available, it would require a high degree of expertise to comprehend it. How do those purchase records and social media posts translate to character traits and political preferences? Because of the much-discussed opacity, or 'black box' nature of some AI systems, even those proficient in computer science may not be able to understand an AI system fully. The process of how AI generates an output is largely self-directed (meaning it generates its own strategies without following strict rules designed by the developers), and difficult or nearly impossible to interpret.

Moreover, the novelty of AI and the fast development of new versions, updates, and improvements make it hard to rely on methods to establish trust that require longer time frames. It just takes a few minutes to check whether Spotify recommendations capture your musical taste. But to know whether the system is competent at making recommendations that affect a person over long time spans, such as career or partner choices, we need long-term data. This data won't be available when the technology is implemented; possibly it will never be available for any specific model at the time of use.

The complex, statistical categorizations made by AI further confuse matters. To make recommendations for, or decisions about, individuals, AI commonly creates groups based on opaque statistical correlations, such as 'probably not suffering from a sexually transmitted disease' or 'likely to be an atheist'.[2] But it often remains unclear to the individual how they ended up in a group, and what they have in common with other members of the group. Moreover, the label they are identified with may appear alien and disconnected from their lived experience because it does not have a corresponding group-identity in real life. Should you feel affiliated with the group of novelty-sweater buyers in which the AI has placed you? It becomes difficult to understand what those characterizations mean, to evaluate the appropriate reaction, or to challenge them. Because it is unclear how the AI system generates user profiles, groups, and recommendations, it is hard to know what to do with this information. Should you trust it, discard it, or fight it?

As long as it is just a song recommendation, or perhaps Google labelling you as a thrill-seeker, this might not be a pressing issue. But in some cases, the stakes are substantially higher. AI is being used in job recruiting and selection, in medical decision-making, and in the justice system to assess recidivism risk or for predictive policing. An example of the latter is the Chicago Strategic Subject List or Heat List implemented by the Chicago Police Department. Between 2012 and 2019, they used an algorithm to list and score people according to their probability of being involved in a homicide or non-fatal shooting—either as a victim or as a perpetrator.[3] Police then approached high-scoring individuals to inform them that they were deemed likely to be involved in a shooting and that they, the police, were keeping an eye on them. What should those individuals do with this information? Should they accept it as a reasonable representation of who they are and what their future holds?

The opacity of AI profiling and decision-making leaves few opportunities for negotiation. On what grounds can you disagree with an opinion that is presented with the authority of scientific fact, computational science, and statistics, without knowing how this opinion has been generated? How would people who objected to being on the Chicago Heat List, and to being described as probable future shooters or gun victims, go about contesting this claim, not just on legal grounds, but to convince friends, family, and ultimately themselves that this is not who they are? The question who really knows you better, yourself or AI, and who has the authority about who you are, remains hard to answer as long as it is unclear whether, in which situations, and to what degree you can trust the AI. In fact, in the case of the Chicago Heat List, a study exposed how ineffective the programme was in combating crime.

Create Yourself!

Even if we had a reasonably trustworthy AI, a second ethical concern would remain. An AI that tells you who you are and what you should do is based on the idea that your identity is something you can discover—information you or an AI may access. Who you really are and what you should do with your life is accessible through statistical analysis, some personal data, and facts about psychology, social institutions, relationships, biology, and economics. But this view misses an important point: we also *choose* who we are. You are not a passive subject to your identity—it is something you actively and dynamically create. You develop, nurture, and shape your identity. This self-creationist facet of identity has been front and centre in existentialist philosophy, as exemplified by Jean-Paul Sartre. Existentialists deny that humans are defined by any predetermined nature or 'essence'. To exist without essence is always to become other than who you are today. We are continually creating ourselves and should do so freely and independently. Within the bounds of certain facts—where you were born, how tall you are, what you said to your friend yesterday—you are radically free and morally required to construct your own identity and define what is meaningful to you. Crucially, the goal is not to unearth the one and only right way to be but to choose your own, individual identity and take responsibility for it.

This self-creation occurs in two, mutually influencing ways: via action and via interpretation. First of all, you decide who you are through your actions. By acting, you create facts about yourself. You become a person who has done X or refused to do Y. Either you are a person who comes to help his friend in need, or you are not.

Moreover, by acting in certain ways and exposing yourself to certain people, emotions, or environments, you gradually shape yourself and change your views, values, beliefs, and/or desires. This might be done in a self-conscious attempt at changing yourself, but we often shape ourselves less purposefully. By choosing a career, for instance, you expose yourself to a certain type of person and certain situations which will influence who you are, whether you like it or not.

At the same time, you also create yourself by conceptualizing and interpreting yourself. The crude facts about people and their actions do not by themselves define their identity. They provide the raw material for self-definition. Figuring out what the facts *mean* is a matter of interpretation, and, to an extent, that interpretation is up to you. Thus, defining yourself involves interpreting actions, understanding what motives drove them and how they connect to overarching intentions, evaluating and organizing personal information, finding patterns, labelling yourself, and making choices about what is important to your identity and what is irrelevant, or which events, social groups, actions, or achievements are defining you and in what way. Self-defining actions and crude facts leave room for creative, interpretive self-definition. To some degree, it is up to you whether you find your teenage celebrity crush embarrassing or embrace it as part of who you are, whether your nationality is a contingent fact about you or something you deeply identify with, or whether doing philosophy research is just a job or a calling. This self-image you create is action-guiding. You get to plan and lead a life in accordance with who you take yourself to be. This interpretive dimension of self-creation feeds back into the agentive one. At the same time, creating yourself through interpretation takes your actions, as well as other facts about yourself, as a starting point.

AI can give you an external, quantified perspective which can act as a mirror and suggest courses of action. But you should stay in charge and make sure that you take responsibility for who you are and how you live your life. An AI might state a lot of facts about you, but it is your job to find out what they mean to *you* and how you let them define you. The same holds for actions. Your actions are not just a way of seeking well-being. Through your actions, you choose what kind of person you are. Blindly following AI entails giving up the freedom to create yourself and renouncing your responsibility for who you are. This would amount to a moral failure.

There is great value in the process of choosing for yourself. The deliberation and experimentation involved in making your own choices is an exercise in self-understanding and self-making. Through this process, you learn and decide what your opinions are or why you like X and identify as Y. You ascribe meaning to actions and facets of your identity when you deliberate and choose because they become connected to insights, beliefs, values, and overarching intentions. You may come to realize that you chose a career because you value the security and stability it provides. Through this process, you also understand the meaning of this job for you, as a source of security. If an AI made this choice on your behalf, the job might not carry this kind of meaning (even though you could reconstruct and ascribe this meaning in retrospect). Moreover, by choosing for yourself, you can be sure that your own interests are guiding the process. When you rely on others or on AI, different interests can come into play. Companies and government institutions that use and provide AI profiling and recommendation systems pursue their own concerns, which might be detrimental to your own well-being. While it might be financially beneficial for YouTube if you watch more of their recommended videos, it could be bad for your well-being if you are thereby politically radicalized and start believing in conspiracy theories.

Ultimately, relying on AI to tell you who you are and what you should do can stunt the skills necessary for independent self-creation. If you constantly use an AI to find the music, career, or political candidate you like, you might eventually forget how to do this yourself. AI may deskill you not just on the professional level but also in the intimately personal pursuit of self-creation. You might profit from using AI as a useful tool to make suggestions for your decision-making and self-interpretation. But given the value of self-creation, you should be careful to retain the skills for doing this on your own. Choosing well in life and construing an identity that is meaningful and makes you happy is an achievement. By subcontracting this power to an AI, you gradually lose responsibility for your life and ultimately for who you are. You no longer deserve blame or praise for your choices and tastes (or, in the extreme case, for how your life evolves) because you are only a passive recipient of choices made by a computer.

Of course, we often make bad choices. But this has an upside. By exposing yourself to influences and environments which are not in perfect alignment with who you currently are you develop. Moving to a city that makes you unhappy could disrupt your usual life rhythms and nudge you, say, into seeking a new hobby. This would change you into someone with novel preferences and routes to well-being. Recommendation systems look at who you are now. Constantly relying on AI recommendation systems might calcify your identity. This is, however, not a necessary feature of recommendation systems. In theory, they could be designed to broaden the user's horizon, instead of maximizing engagement by showing customers what they already like. In practice, that's not how they function.

This calcifying effect is reinforced when AI profiling becomes a self-fulfilling prophecy. It can slowly turn you into what the AI predicted you to be and perpetuate whatever characteristics the AI picked up. By recommending products and showing ads, news, and other content, you become more likely to consume, think, and act in the way the AI system initially considered suitable for you. The technology can gradually influence you such that you evolve into who it took you to originally be. If Google thinks you like SUVs and bombards you with SUV ads and content, you are more likely to develop a preference for SUVs. Because recommendation systems perpetuate identified characteristics and those characteristics are often modelled after patterns found in other people—what does someone want who is your age, who visits the kind of locations you regularly go to, and who has your type of purchase record and news consumption—they level out outliers. While those recommendations aim to be individualized and uniquely customized for you, they often build on universals and averages, on data about what others with your characteristics do, want, and choose (this is called collaborative filtering). They do not make everyone the same, but they do make you more likely to resemble others within your specific bubbles.

The Chicago Heat List became a self-fulfilling prophecy for at least one person on the list.[4] Robert McDaniel had no felonies or violent offences on his criminal record. He had no idea why he was deemed likely to be involved in a shooting or how he could change that. McDaniel was repeatedly visited by the police and was put under constant surveillance after the algorithm gave him a high score on the Heat List. As a result, people who noticed the regular police interactions suspected him to be an informant and shot at him on two separate occasions. He was injured but survived both incidents.

*

You may sometimes wish for someone to tell you what to do or who you are. But, as we have seen, this comes at a cost. It is hard to know whether or when to trust AI profiling and recommendation systems. More importantly, by subcontracting decisions to AI, you may fail to meet the moral demand to create yourself and take responsibility for who you are. In the process, you may lose skills for self-creation, calcify your identity, and cede power over your identity to companies and government. Those concerns weigh particularly heavy in cases involving the most substantial decisions and features of your identity. But even in more mundane cases, it would be good to put recommendation systems aside from time to time, and to be more active and creative in selecting movies, music, books, or news. This in turn, calls for research, risk, and self-reflection.

Notes

1. Y. N. Harari, *Homo Deus: A Brief History of Tomorrow* (Harvill Secker, 2016).
2. K. de Vries, 'Identity, Profiling Algorithms and a World of Ambient Intelligence', *Ethics and Information Technology* 12/1 (2010): 71–85.

3. A major problem with the use of AI in the justice system is that it can reinforce or lead to discrimination. For more on this see Binesh Hass's chapter in this volume.
4. M. Stroud, 'Heat Listed', *The Verge* (2021), <https://www.theverge.com/c/22444020/chicago-pd-predictive-policing-heat-list>.

Further Reading

J. Cheney-Lippold, *We Are Data: Algorithms and the Making of Our Digital Selves* (New York University Press, 2017).

16
Information Flows in the Digital Age

Emma Bluemke and Andrew Trask

'Information Flows'

What's private to you?

Well, it depends, right? Private from whom?

What would you tell a family member, but not your employer? What would you share with a researcher but not an app? What story would you reveal online anonymously—maybe on Mumsnet or a Reddit forum—but you wouldn't even reveal to a friend?

Is my face private? You can see mine online right now if you search.

But if my face is public information, why does facial recognition technology make me uncomfortable? If my location is known to anyone standing near me on the street, why am I uncomfortable with having my location tracked? What's the difference?

The simple word 'privacy' gets complicated quickly. Privacy doesn't seem to be about the information itself, but about the way it flows. This begs the question: 'why?' Why are we comfortable with sharing a bit of information in *some* contexts, but not others? Where is that line drawn? When do people consider shared information to be a privacy invasion?

To answer these questions, privacy philosopher Helen Nissenbaum introduced the concept of 'contextual integrity'. Her aim was to pinpoint exactly when an information flow crosses this invisible line. In one example, she looked into why people were so upset about photographs of them being captured by Google StreetView: after all, they're standing out in public, right? But Nissenbaum pointed out that when you take this public information and record it, store it, and make it available in unexpected ways, it becomes part of a very different *information flow.*

A great example of a privacy standard being about the *flow* and not the *information itself* is facial recognition. We all consider our faces to be largely public information, and we reveal them every day when going about day-to-day life. But facial recognition software as a form of ID has important differences from normal identification. First, it's difficult, if not impossible, to change your face. Just like your DNA or your fingerprint, if it gets into the wrong hands, you can't just 'set a new password' or 'order a new credit card' and be safe. Second, unlike other forms of identification, facial recognition's primary feature is that it is identification without your consent. This means the information flow for identification is not triggered by you, but by whatever system is watching you. This violates people's sense of privacy not because a face is private information per se, but because the information flow from facial images could be used to cause harm—such as supporting a mass surveillance programme.

Location data is similar. Despite the fact that your location is obvious to people around you, revealing your location can also be a violation of privacy if done in the wrong context.

The word 'privacy' often gets a bad rap—people equate it with secrecy, or locking down information. But, with these examples, it's clear that the same information might be considered private in one context, but not in another. Or, more specifically, sharing information in one context might seem like a privacy invasion, while sharing that same information in another context might be totally acceptable. Privacy is not about hiding or secrecy—it's about the information flow, and our focus should be on achieving appropriate information flows.

Across social contexts such as education, healthcare, and politics, societies have developed norms that regulate the flow of personal information. These norms help protect individuals and groups from harm and help balance power among competing actors. However, in a globalised world where much information is stored digitally in forms that allow easy copying, sale, and exploitation, enforcing our old context-relative norms is becoming increasingly challenging.

The fact that data can be stored for decades makes these norms even more difficult to enforce—what might this information reveal about us in 10 years' time? There are various concerns.

First, as social norms and political regimes change over time, a fact that I'm currently comfortable sharing (e.g. my religious affiliation) may be used against me. Second, changes in my data over time can reveal more about me than I realise—my grocery bill doesn't feel private today, but when compared with my bills from six months ago it may suggest I'm pregnant or ill. Third, our ability to infer and discover patterns in data is only increasing—for example, my brain MRI is in a public research dataset; it's currently anonymised and seems innocuous, but in 20 years, will I realise that I've accidentally revealed my likelihood of developing dementia to my medical insurance provider?

Today, society is actively working out the question: what *is* an appropriate flow of information in this new digital age? What's appropriate for any given context, and how do we achieve and enforce that?

Consider the following uses of information:

- We want diseases to be detected and cured—but we don't want our medical information to be misused.
- We want every vote to be counted and to have trust in the democratic process—but we want votes to be anonymous.
- We want to be able to video-chat with our closest friends and family members across the ocean—and ask them deeply personal questions—but we don't want the people running the communication infrastructure to use that information against us.
- We want people who cause serious harm in society to be stopped—but we don't want a surveillance state.
- We want to allow access to monitor biases and other harmful impacts of algorithms—but don't necessarily want to reveal the algorithm publicly for anyone in society to misuse.

In each of these situations, we know of a very specific flow of information—restricted to very specific people—that could accomplish our shared goal. How might we build these information flows?

Balancing Privacy, Accountability, and Transparency

These scenarios demonstrate the tensions that exist between transparency, privacy, and accountability in the use of data and algorithms. At its core, the issue is allowing the appropriate use of information while preventing its inappropriate use. We call this goal 'structured transparency'—bringing the benefits of transparency, but in a structured way. The goal of structured transparency is to enable precise social or technical arrangements that determine *who* can see *what*, *when*, and *how*.

Although the term 'structured transparency' is relatively new, the goal is ancient. For example: the ancient practice of casting secret ballots allows voter preferences to be expressed without intimidation and shaming. That's structured transparency. A modern example is sniffer dogs in airports: detection dogs report that a bag is safe without anybody having to rummage through its contents. They can alert us if there are drugs, bombs, or other contraband in a suitcase, without revealing our dirty old socks. That too is structured transparency.

Core Challenges

Structured transparency is about achieving a desired information flow, answering the questions: who should be able to know what, when should they be able to know it, and what should they be able to do with this knowledge?

If we break it down, there are a few key problems that make it difficult to enforce who can see what, when, and how:

The copy problem: The copy problem arises when we share information with others; this makes it challenging to regulate how they use or disseminate it. Once we give someone a piece of information, we lose control over its fate. This is especially true with digital information, which is easy to copy and distribute. This can be particularly important when sensitive information is involved, as it can lead to unintended consequences such as privacy breaches and discrimination. For instance, if we share our health details with an app, we expect this information to be used only for the purpose of helping us, but there are no technical (as opposed to legal) guarantees that prevent the app from sharing it with others (e.g. health insurance agencies) or using it for unrelated purposes (e.g. targeting us with weight loss advertisements).

The bundling problem: The bundling problem occurs when we need to share certain information, but it's tied up with other details that we don't want to disclose. For example, when you show your driver's licence to prove you're old enough to enter a club, you also reveal irrelevant personal information like your name and address. In order to balance the need for verification with privacy concerns, we need to find ways to untangle the information we want to share from the information we want to keep private.

The edit problem: Imagine you've sent an important document to someone, only to find out later that they made changes to it before forwarding it to others. This same issue can arise with information stored and transmitted electronically, including with AI algorithms. If a company makes edits to an algorithm after it has been audited and approved, it would be difficult for me, the customer who is receiving the recommendations or output, to know that the algorithm I'm receiving has been altered from its approved version. This has important implications for ensuring responsible use of AI in sensitive applications: for example, someone building an education app that incorporates a large language model would need to have an absolute guarantee that they're receiving a version of the AI model that has been screened and approved for interacting with children.

Using third-party organisations to oversee data sharing and AI models can help address issues with controlling information use, but it can also create another problem:

The recursive oversight problem: By adding a third party to oversee data sharing, we're creating another entity that could potentially misuse that information. This raises an important question: how can we make sure that the third-party organisation itself is trustworthy and accountable? In other words, 'who's watching the watchers'?

New Tools

Underneath all these problems is the trade-off between privacy and transparency. However, new privacy-enhancing technologies (PETs) are working to solve that trade-off in different ways. A lot of these technologies sound like alphabet soup: homomorphic encryption, differential privacy, or zero-knowledge proofs. Very briefly, here's a simplified overview of some of the most popular techniques:

- *Federated learning* means training your machine learning algorithm on data that is stored on different devices or servers across the world, rather than it all being collected from one place. For example, an algorithm could be trained on data from several hospitals without needing to move the private medical information from the hospitals' servers.
- *Encrypted computation* techniques allow machine learning on data that remains 'secret'. For example:
  - *Homomorphic encryption* makes data unreadable yet still allows maths to be performed on it (e.g. to answer research questions). For example, the finance industry could use this to catch patterns of money laundering in data shared across borders without violating the privacy of their customers.
  - *Secure multi-party computation* allows multiple parties to collaborate together while keeping the inputs and the algorithm itself private (e.g. it allows an algorithm to be trained (or used) on data from different sources while keeping the data or algorithm 'secret'). For example, a facial recognition algorithm could be jointly governed by human rights and

law enforcement organisations to ensure the algorithm is only used when the parties agree to the use case.

- *Differential privacy* allows us to study the patterns within the dataset while protecting information about individuals in the dataset. For example, national census data can be collected and analysed in a way that provides accurate group-level statistics without revealing specific traits about individual people.

However, explaining what these technologies do doesn't tell you what they're good for. Which of these technologies do the same thing? Where do they overlap and is that overlap total or partial? Which can be used together? Which industries will they affect? Who should be the first users? And most importantly, will this technology actually make our society better?

The following structured transparency framework might help. It can show how a particular information flow you're interested in creating can be broken down into the individual challenges necessary to accomplish that data flow: input privacy, output privacy, input verification, output verification, and flow governance. Then, we can figure out which of the many existing technologies could help achieve that flow.

Input privacy techniques allow us to create a flow of information between multiple people while keeping the value of each person's input secret. For example, these techniques allow researchers to use medical data to answer their research question without being able to *see* that private information themselves.

Input privacy is the most common focus of discussions around 'data privacy', but often it's not enough. Sometimes, AI models can 'memorise' details about the data they've trained on and can 'leak' these details later on—for example, a language model can be probed to release information about individual samples that were used for training (e.g. individual peoples' names, phone numbers, addresses, and social media accounts). This is why output privacy is important:

Output privacy techniques prevent the algorithm's output from being reverse engineered to infer harmful unintended information about the input (or the sender of an input).

At first glance, satisfying input and output privacy seems enough to successfully achieve an information flow. However, in practice, the act of hiding all but the information you want to transfer can raise doubts about the final result. For example, if an organisation keeps its training data private, how can researchers verify traits about the underlying training data (e.g. copyright status, bias)? This is why we often need input and output *verification* as well.

Input verification techniques allow a party to verify specific attributes of an input to an information flow, for example that it came from a trusted source or that it happened within a specific date range. These techniques may become important for governing AI use: for example, developers of apps for education or entertainment may need to have guarantees that they're receiving a version of the algorithm that is safe for children.

Output verification is concerned with verifying attributes about what's happening during the hidden flow itself. When combined with input privacy techniques, tools for output verification can overcome the recursive oversight problem. For example, an external auditor could verify properties of an information flow without learning anything beyond the output of targeted tests (e.g. searching for patterns reflective of fraud) while also ensuring that the tests ran correctly. This has significant implications for increasing the precision, scale, and security of auditing institutions, potentially facilitating new types of checks and balances and fairer distributions of power.

Finally, even if some subset of input and output privacy and verification is satisfied by an information flow, there remains the issue of who has the authority to modify the information flow itself. This is where flow governance comes in:

Flow governance is satisfied if each party with concern/standing over how information should be used has guarantees that the information flow will adhere to the intended use. A non-technical analogy for these techniques is multi-key safety-deposit boxes for holding secure documents.

This framework shows what privacy-enhancing technologies are capable of doing—without you having to understand the technologies yourself. Part of its purpose is to build a bridge between technical and non-technical communities so that non-technical folk know what problems and tools they have and—inversely—so that technical folk know what end-goals they're supposed to accomplish with privacy technologies.

Example 1: Improved Governance of Data Use

Input privacy techniques could enable data owners (hospitals, labs, statistics offices, etc.) to grant researchers the ability to perform specific computations over their data without providing access for any other operations. This would allow the researcher to answer their research question ('what is the mean weight of newborns in this city?') without the data owner having to worry about the copy problem. Additionally, using output privacy techniques, data owners could prevent reverse engineering of the computation output to reconstruct patient information.

Input verification techniques could allow data owners to prove various attributes of the dataset to researchers—such as the demographic distribution. In situations that feature a competitive relationship between institutions or research groups, output verification may also be required, to prove, say, that a key statistical result was reached from computations requested by the researcher (as opposed to stemming from shortcuts or mistakes).

Finally, flow governance could distribute control across third parties (funding bodies, stakeholders in a collaboration, groups safeguarding rights for vulnerable populations, etc.) to enable sensitive information to remain available for appropriate research while minimising risk of misuse. In summary, stronger, more precise, and more automated controls over data sharing could make more scientific data available for research or public collaboration, increasing the pace of research in many fields.

Example 2: Improved Governance of AI Use

Input privacy techniques could enable algorithm owners to grant researchers the ability to perform specific computations over their algorithm (e.g. evaluating bias) without providing access for any other operations. Additionally, output privacy techniques could stop the algorithm from being reverse engineered.

Input verification techniques could allow algorithm owners to prove various attributes of the algorithm to researchers—such as the source of the underlying data that the algorithm was trained on. In addition, it may be possible to 'watermark' some information in order to reliably check whether a piece of information was generated by an algorithm or, vice versa, check that it was generated by a human.

In situations that feature a competitive relationship between institutions or research groups, output verification could help check that the result of the evaluation was computed by the algorithm owner using the exact computations requested by the researcher or auditor (i.e. showing that the algorithm owner did not alter any of the evaluation code).

In some cases, it may be desirable for an AI algorithm to be governed by multiple parties. For example, if multiple actors bear the cost of creation by pooling their datasets they may wish to ensure that subsequent commercialisation of the asset distributes profits amongst the group. Flow governance could distribute control across third parties (e.g. funding bodies, watchdog organisations), thereby making especially sensitive models available for appropriate research only when approved by several parties.

What Does This Change?

These techniques provide two important new abilities. First, the potential for information to be unbundled so that only necessary information for a collaboration is shared. Second, the breaking of the recursive enforcement problem so that small actors can audit information to which they themselves don't have access.

When used together, the various aspects of structured transparency allow us to benefit from the *desired* uses of information without enabling *misuse*. These tools provide the infrastructure to create precise social and technical arrangements that determine who can see what, when, and how.

While more technology alone will not entirely resolve all social information-sharing issues, technical tools for structured transparency are promising because they allow us to demand that companies abide by certain information flows. For example, if these technologies make it possible for a mobile app (maps, health apps, etc.) to provide their service without aggregating and storing your location or health information on their servers, then we (and regulators) can insist that the app uses this new technology. Structured transparency tools change what it's possible to demand.

What Can I Do?

It can be overwhelming to hear the many ethical issues surrounding AI and data use. However, we want this chapter to give you optimism—to feel inspiration, ambition, and hope that it's possible to design, create, and enforce much better information flows than we have today.

Not only can we achieve better information flows—we *must* achieve better information flows if we want to avoid undesirable outcomes in a future with multiple systems using our data and AI affecting many areas of our lives.

But the hardest part of achieving that better future isn't technical. The hardest part is figuring out how society can shift gradually from where we are now to a system with structured transparency—we need to combat all sorts of obstacles, including the network effects and legacy infrastructure of our current systems.

But these obstacles must be overcome. We must start with the question of what is the most ideal information flow for this context. As a consumer, start to demand that. As a community, start to arrange that. As a researcher or engineer, start to build that. As a business, start to guarantee that. As a lawmaker or regulator, start to enforce that. And, as an investor or grant-maker, invest in the infrastructure needed to create a society we all want to live in.

Although many of these technologies are already in use, some are still under development, and researchers are actively working to invent faster versions or ways to implement them practically. However, just because a technique is 5–10 years away from mainstream use doesn't mean you should dismiss it—a platform for instant video calling was science fiction a few decades ago, yet it's now a part of our everyday lives. If we want those cutting-edge techniques to exist, it's a research opportunity and an opportunity for government investment. If we want existing techniques to be more widely applied, it's an opportunity for entrepreneurs, investors, activists, and lawmakers. And if we want techniques that haven't yet been invented, that's for the science fiction authors—and all of us—to daydream.

Further Reading

Emma Bluemke and Andrew Trask, *Private AI Series, Course 1: Our Privacy Opportunity*. Published by OpenMined in partnership with PyTorch, University of Oxford, and United Nations Global Working Group on Big Data. 2021. <https://courses.openmined.org/courses/our-privacy-opportunity>.

PART V
IDENTITY AND VALUES

17
Robotic Persons and Asimov's Three Laws of Robotics

César Palacios-González

Introduction

Imagine that Ana and Rob are the only passengers in a plane that is going to crash, through no fault of theirs. Because there is only one parachute one of them is going to cease to exist. The presumption must be that both individuals have an equal claim to the parachute. Unfortunately for Rob, he is a robot—and has been programmed with the Three Laws of Robotics (TLR), the first of which maintains that: a robot may not injure a human being, or, through inaction, allow a human being to come to harm. Rob should better start preparing for robotic afterlife. In this chapter I will explore why robotic persons pose a problem for the TLR, and how the problem can be dealt with.

The Three Laws of Robotics

The novelist and polymath Isaac Asimov was fascinated by the literary potential of robots, but irritated by the overuse of 'Frankenstein Complex' plot lines, in which a human creates a robot which then turns on its creator. In order to sidestep this trope he came up with a set of hierarchical laws which would prevent, or dramatically reduce, 'robot kills human' scenarios.[1] Robots, in his imaginings, would not be able to contravene the laws. The TLR appeared explicitly for the first time in the 1942 short story 'Runaround'.[2] The three laws are:

1. A robot may not injure a human being, or, through inaction, allow a human being to come to harm.
2. A robot must obey the orders given it by human beings except where such orders would conflict with the First Law.
3. A robot must protect its own existence as long as such protection does not conflict with the First or Second Law.

In the story 'Robots and Empire',[3] Asimov revisited the laws and introduced a zeroth law. This allows robots to cope with situations in which they had to harm, or kill, human beings in order for a larger number of human beings to remain safe. The zeroth law states:

- A robot may not injure humanity, or, through inaction, allow humanity to come to harm.

Since their appearance the laws have become an inescapable part of popular culture, appearing, for example, in *The Simpsons*, *Doctor Who*, and the *Alien* franchise.

The Place of the Three Laws of Robotics

It might be thought that the TLR were conceived *only* as a literary device to help Asimov explore fictional scenarios. Not so. Asimov believed that his TLR were going to play an important part in robotic research and development.

> I have my answer ready whenever someone asks me if I think that my Three Laws of Robotics will actually be used to govern the behavior of robots, once they become versatile and flexible enough to be able to choose among different courses of behavior. My answer is, 'Yes, the Three Laws are the only way in which rational human beings can deal with robots—or with anything else.'[4]

In fact, the TLR have not become integral to robotics or artificial intelligence. As computer scientists Robin Murphy and David D. Woods have shown, there is almost no academic discussion on how to code and implement the TLR, or even if they represent a viable framework for robot–human interaction.[5]

The technologist Roger Clarke[6] concluded that Asimov's own science-fiction literary work shows that the TLR are not foolproof. They don't prevent humans being harmed by robots because in any given situation there could be: (a) ambiguity and cultural variation of terms (e.g. what does 'harm' mean), (b) a role for judgement in decision making (e.g. how to prioritise the received orders), (c) sheer complexity (e.g. how to navigate a situation where there are many unknown factors), and (d) room for dilemma and deadlock (e.g. what to do when two contradictory orders are given at the same time).

In a similar vein, Oxford philosopher Nick Bostrom maintains that the main issues with the TLR are definitional (e.g. what is harm) and 'vagueness'.

> Consider, for example, how one might explicate Asimov's first law. Does it mean that the robot should minimize the probability of any human being coming to harm? In that case, the other laws become otiose since it is always possible for the AI to take some action that would have at least some microscopic effect on the probability of a human being coming to harm.[7]

Other philosophers have pointed out different problems—for example, that a robot might not understand the consequences of an action (i.e. that it is violating the TLR),[8] and that there are *practical* shortcomings of the TLR: (a) they are anachronistic (e.g. they do not take into account the weaponisation of robots), (b) they leave the robot in charge of safety-related matters when interacting with humans, (c) AI robust natural language understanding is still under development, (d) communication is not only verbal, and (e) the third law is inadequate in that robots should not be programmed to follow, indiscriminately, orders given by *any* human.[9] All these criticisms succeed in highlighting problems with the *applicability* of the TLR in both fictional and non-fictional worlds.

These criticisms are well known. But here I want to explore a question that has for the most part been overlooked. Which robots, if any, should be programmed with the TLR?

It seems likely that this question has been overlooked because those writing about the TLR have assumed they should be programmed in *all robots*—I call this position the 'Every-Bot-TLR'. But it's an assumption worth interrogating.

The Every-Bot-TLR position interprets the formulation—'A robot must' or 'A robot may'—to mean *all* and *any* robots. Through almost all of his work Asimov embraced this position, with one notable exception which I will explore later. But the Every-Bot-TLR position is morally problematic, because it misses out the morally relevant differences that could exist between different types of robots.

Plane Crash

Before proceeding with some thought experiments let me clarify a term I will use, a 'person'. I use this term in a technical sense, to mean 'a thinking intelligent being, that has reason and reflection, and can consider itself as itself, the same thinking thing in different times and places'.[10]

Imagine the following three cases:

1) Ana and Mary (two adult women) are the only passengers in a plane that is going to crash, through no fault of their own. Because there is only one parachute one of them will die.

2) Ana (an adult woman) and B-9 (a non-person artificial intelligent robot) are the only passengers in a plane that is going to crash, through no fault of their own. Because there is only one parachute one of them will soon cease to exist.

3) Ana (an adult woman) and Rob (a robotic person) are the only passengers in a plane that is going to crash, through no fault of their own. Because there is only one parachute one of them will soon cease to exist.

Who should get the parachute and why? In normal circumstances, other things being equal, Ana and Mary must be considered equally valuable from an ethical point of view. Ana's interest in remaining alive does not trump Mary's interest in remaining alive, nor vice versa. Both women have the same claim to the parachute, irrespective of the fact that inevitably one of them will die.

Case (2) is just as easy. If Ana is a person and B-9 is a non-person artificial intelligence, then Ana's moral status trumps B-9's existence. Ana has a stronger claim to the parachute because she has a higher moral status than B-9. It is true that B-9 might have some degree of moral status, for example one comparable to that of a non-person non-human animal (such as a goat), but even then B-9 should not get the parachute, just as in normal circumstances we would not accept that a goat has a stronger claim to the parachute than a normal adult human being.

Case (3) is the most interesting. Here, Ana is a biological human person and Rob is a non-biological robotic person. This means that Ana's and Rob's moral status is equivalent. Both of them possess the cognitive capacities that grant personhood. If the *possession* of such cognitive capacities is what grants personhood then the fact that their substrates are different (Ana's substrate is biological while Rob's substrate is non-biological) is irrelevant from a moral point of view. We can thus conclude that, as in case (1), *both* of them have the same claim to the parachute.

Obviously, there are serious questions about the feasibility of developing true human-like artificial general intelligence.[11] Here I will simply bracket this discussion and assume: (i) that Rob is a human-designed artificial general intelligence with the same capacities as any other normal human adult and (ii) that the only difference between Ana and Rob is that one type of person occurs 'naturally' and the other is developed 'artificially'.

Now, suppose that Rob and Ana have the same claim to the parachute. What would actually happen if Rob was programmed with the TLR?

Well, first, Rob would not be able to compete with Ana for the parachute because it would break the first law, in that by an *action* (fighting over the parachute) or *inaction* (leaving Ana in the doomed plane) Rob would allow a human being to come to harm.

Second, even if Ana wanted to give Rob the parachute voluntarily, Rob would not be able to use it because that would also violate the first law; by taking the parachute Rob would allow a human being to come to harm.

Third, if Ana wanted to stay on the plane with Rob (knowing that Rob would not be able to receive or 'fight' for the parachute), Rob would have to save Ana even against her will; not to do so would violate the first law in that by *inaction* (i.e. not forcing Ana to jump with the parachute) would allow a human being to come to harm.[12]

Fourth, if Ana wanted to save Rob and she ordered him to put the parachute on, Rob would have to disobey and force Ana to take the parachute instead. The second law states that robots must obey human orders, but in this case such an order conflicts with the first law, which it cannot violate.

Finally, even if Rob appealed to the third law, which mandates that he must protect his existence (and so, presumably, 'fight' Ana for the parachute), this conflicts with the first law, which has primacy. In any case, Ana could simply command Rob to give her the parachute and Rob would be unable to refuse.

So in this scenario, Rob's survival would be impossible, even though he has the same moral claim to the parachute as Ana. The TLR forces Rob to act as if he were *merely an object* to be sacrificed for Ana's sake, despite having a moral status that entails, in Kantian terms, that he should be treated as an end in himself.

Not All Robots Are Equals

The main problem with the Every-Bot-TLR position is that it fails to take into account that in the future there could be robotic persons. Programming a robotic person with the TLR would fail to respect their moral status in at least four ways.

First, it would impose an external constraint upon them that substantially restricts their ways of interacting with the world. Second, it would make it impossible for robotic persons to follow freely chosen courses of action. Third, under certain circumstances the TLR would force robotic persons to act against their own fundamental interests. Fourth, robotic persons programmed with the TLR are de facto slaves to humans, since they cannot refuse to comply with their orders.

Programming the TLR into Rob is morally on a par with implanting a brain chip in a normal adult human, against her will, that would make her comply with the TLR. Imagine this modified version of the first scenario.

Ana (a normal woman) and Shelly (a normal woman who against her will has been implanted with a brain chip that makes her abide by the TLR) are the only passengers in a plane that is going to crash, through no fault of their own. Because there is only one parachute one of them is going to die.

In this case Shelly will inevitably die (for the same reasons as Rob will inevitably cease to exist). Now, if we think that programming Shelly with the TLR is immoral, because her moral status demands otherwise then, to be morally consistent, we need to maintain that programming Rob with the TLR is also immoral.

How would we respond to someone who, faced with this argument, nonetheless continued to maintain the Every-Bot-TLR position? Just as in the past people have relied on *non-morally relevant* features (like nationality, gender, or skin colour) to maintain that certain groups possess superior moral status, it is not far-fetched to suppose that in the future some people will point to the fact that human beings are 'made of flesh and bones' in order to claim that robotic persons are less valuable—and to justify programming them with the TLR. This attitude would be an instance of what can be called 'Bioism': the prejudice or bias in favour of biological entities with X interests and capacities over those of non-biological entities with equivalent X interests and capacities. And just as we reject racist, sexist, or speciesist attitudes because they focus on the wrong features to extract normative conclusions, we should also reject bioism, because it picks out a non-morally relevant feature of human persons (their biological substrate) to the detriment of robotic persons.

Robotisation

Another way to show the immorality of the Every-Bot-TLR position is with a second thought experiment: 'robotisation'. Suppose that over the course of a year a human person, because of an illness, swaps her biological body for a robotic one. Let's suppose too, that her cognitive capacities and personal identity survive such a change. Let's further suppose that the replacement happens gradually, first an arm, then a leg, and so forth until her biological brain is replaced (neuron by neuron) with a robotic one. If we were to abide by the Every-Bot-TLR position then we would have to accept that at some point of the 'robotic transformation', the point at which she becomes a robot, she should be programmed with the TLR. My guess is that those who hold that a human person should not be ruled by the TLR would not agree with the TLR being imposed into this biological-to-robotic person. And that's because they would recognise that the robot possesses the same cognitive capacities as before and there is no break in personal identity. The human person's identity appears to survive the robotisation process.

To complicate matters further, now imagine that a robotic person, programmed with the TLR, decides to undertake the inverse operation in the exact same time lapse. Little by little her robotic body is replaced with a biological one, until a full transformation is achieved. If this were to be the case then at some point, the point at which she *is not* a robot, the TLR should be removed.

'Robotisation' shows that either we accept that we should impose the TLR in the biological-to-robotic person cases and remove the TLR in the robotic-to-biological person cases, or we recognise that imposing the TLR into robotic persons is morally wrong because robotic persons possess the same relevant set of capacities as biological persons—capacities that outlaw the restrictions that the TLR impose.

A Narrow Reading of the Three Laws of Robotics

This conclusion leaves us with two options. Either we discard the TLR completely[13] or we postulate a principle that recognises the morally relevant distinctions between robotic persons and robotic non-persons. A simple principle is this:

The Three Laws of Robotics should not be programmed into robotic persons.

By narrowing the scope of TLR implementation we avoid the main ethical problem (that of robotic persons being programmed with the TLR) that follows from the Every-Bot-TLR position. By not programming the TLR into robotic persons we respect their moral status, while at the same time cashing in all the benefits that derive from the implementation of the TLR into robotic non-persons.

Would Asimov have embraced this narrow implementation of the TLR? It's difficult to be sure because it opens up the possibility of the human-creates-robot, robot-turns-on-its-creator scenario. However, in his novelette *The Bicentennial Man* Asimov features a robotic person, Andrew Martin. The TLR cause Martin all sorts of problems. It's fair to assume that this story reflects Asimov's realisation that the TLR are indeed morally problematic.

Notes

1. Asimov is considered the creator of the TLR. However, he gave the credit to John W. Campbell, who was the editor of the *Astounding Science Fiction* magazine. Campbell denied this.
2. Isaac Asimov, *I, Robot*, reprint edition (Spectra, 2008).
3. Isaac Asimov, *Robots and Empire* (Doubleday & Company, 1985).
4. Isaac Asimov, 'The Three Laws', *Compute!* (November 1981): 18.
5. R. R. Murphy and D. D. Woods, 'Beyond Asimov: The Three Laws of Responsible Robotics', *IEEE Intelligent Systems* 24/4 (2009): 14–20, doi:10.1109/MIS.2009.69.
6. R. Clarke, 'Asimov's Laws of Robotics: Implications for Information Technology, Part 1', *Computer* 26/12 (1993): 53–61, doi:10.1109/2.247652; R. Clarke, 'Asimov's Laws of Robotics: Implications for Information Technology, Part 2', *Computer* 27/1 (1994): 57–66, doi:10.1109/2.248881.
7. Nick Bostrom, *Superintelligence: Paths, Dangers, Strategies* (Oxford University Press, 2014), 139.
8. See Lee McCauley, 'AI Armageddon and the Three Laws of Robotics', *Ethics and Information Technology* 9/2 (2007): 153–64, doi:10.1007/s10676-007-9138-2.

9. See Murphy and Woods, 'Beyond Asimov'.
10. John Locke, *An Essay Concerning Human Understanding*, ed. Peter H. Nidditch (Clarendon Press, 1979), Book II, chap. 27, sec. 9.
11. Throughout this chapter I talk about robots, but my arguments work, other things being equal, for any other type of AGI (Artificial General Intelligence).
12. It might be the case that saving Ana against her wishes also harms her; here, however, the harm of dying (and the process of dying) seems to be the greater harm.
13. As Susan Leigh Anderson recommends.

Further Reading

Susan Anderson, 'The Unacceptability of Asimov's Three Laws of Robotics as a Basis for Machine Ethics', in M. Anderson and S. Anderson (eds), *Machine Ethics* (Cambridge University Press, 2011).

18

Is AI Ethics All Fluff?

John Zerilli

> ...among Silicon Valley's princelings, any question that does not yield a verifiable and quantifiable answer is deemed stupid...If it's not maths, it's not real. Society and its expectations about behaviour are lumped together with literature and Santa Claus. Humanities are dumb because you can't have a 'proof'.
>
> —Ed Smith, 'How Michael Lewis Fell for Sam Bankman-Fried,' *New Statesman* (10 October 2023).

Rumour has it that all this talk of AI ethics is much ado about nothing. The concern isn't necessarily—at least not openly—that ethics is useless: it's that what passes for AI ethics is useless. We can all agree that people should be nice to one another; but so much of AI ethics appears to be no more interesting than getting AI to be nice, as it were. If it's wrong to be racist and sexist and homophobic, it's wrong to be these things with algorithms. Is there anything more to AI ethics than getting AI to comport with such rudimentary ideas; and isn't this task anyway a technical matter rather than an ethical one?

The AI ethicist might counter that, while we can all agree that people should be nice to one another, we very often disagree on the details. Being nice to a close friend either means telling them the truth about how they look in that get-up, or telling them they look fine when they don't. It all depends on the occasion your friend is getting dressed up for and whether there's a dress code they're likely to flout, whether they're abnormally sensitive to criticism, whether they're suffering from clinical depression, and so on. Similarly, we can all agree that it's wrong to be racist. Even many racist people would agree. The truly thorny discussions about race aren't to be had over the abstract *ideal* that we should eliminate racism but instead over what *counts* as racism—and very often over forms of racism that are likely to fall beneath most people's radar. Moreover, the AI ethicist can agree that so long as statements of AI ethics principles go no further than platitudes (e.g. AI shouldn't be racist, AI shouldn't breach people's privacy, AI should be transparent and accountable, etc.), then yes, they're next to useless. But this is just to say that AI ethics *properly conceived* is far from useless. It is contentious and consequential and interesting.

Still, this response is unlikely to satisfy the committed sceptic, and there are two such sceptics I have in mind. One I'll call the *philosophical* sceptic insofar as the person in question is likely to *be* a philosopher. 'Sure, there's such a thing as moral disagreement,' they'll say, 'but that's life. There's nothing new to see here, except that instead of these tragic conflicts playing out in a supermarket or on a Broadway stage, they're being played out on a government website or in an online store.' As it happens, I think there are *some* philosophical questions that AI poses for both moral and political philosophy that *are* new (or at least whose resolution has renewed impetus), but that's not what this chapter is about. What I'm interested in counteracting is the second kind of sceptic, whom I'll call the *STEM* sceptic, insofar as the person in question is likely to be someone working in science, technology, engineering, and mathematics (i.e. a 'STEM' field). The reason for this focus is that the STEM sceptic is both far more prevalent than the philosophical sceptic as well as far more pernicious in the influence they are apt to wield.

You might have encountered a STEM sceptic before. Probably the most distinctive mark of the STEM sceptic is a certain dismissive attitude towards non-STEM subjects and non-STEM folk *generally*. I'm talking here about subjects falling within what are called the 'arts and humanities'—history, philosophy, classics, fine arts, design, literature, and so on. Depending on the STEM sceptic you're dealing with, the attitude might even extend to some social sciences like economics and sociology. The attitude is generally expressed in the conviction that all these non-STEM subjects are *soft* subjects, indulged by people without the smarts to do *hard* subjects. Real intellectual clout, they will say, belongs to the STEM student, who can tolerate the demands of a university course in chemical engineering (or statistical physics, or informatics, etc.). On ethics—and AI ethics by extension—well, it's all fluff. And that there's often so much disagreement about how to resolve ethical conundrums proves the point. When a subject is *hard*, there's a right and a wrong answer.

But the humanities folk aren't quite above reproach either. An equally if not more dismissive and even sneering attitude is sometimes found among non-STEM folk toward the STEM folk. They look on the tech world as culturally illiterate, even in some respects daft, holding uncritical and unreconstructed views of human nature, society, gender norms, and politics. There's a reason, after all, why countries governed by totalitarian regimes aren't known for their vibrant theatre districts but can meanwhile churn out STEM graduates like there's no tomorrow. The latter pose far less of a political threat (and give a demonstrable leg-up besides, be it military or economic). Pondering the technological sophistication of the Nazis, George Orwell once described it as science in service of the Stone Age.[1] History shows that the two—science and repressive politics—are certainly compatible, at least for a time.

C. P. Snow was another twentieth-century British novelist, but unlike Orwell, Snow also happened to be an important scientist, so was able to speak with unusual authority about the conflict between what he termed 'the two cultures'.[2] Between the 'literary intellectuals' at one pole, and scientists—primarily physicists—at the other, he found 'a gulf of mutual incomprehension—sometimes...hostility and dislike, but most of all lack of understanding. They have a curious distorted image of each other. Their attitudes are so different that, even on the level of emotion, they can't find much common ground.' Snow goes on to say that the literary intellectuals 'have a rooted impression that the scientists are shallowly optimistic, unaware of man's condition. On the other hand, the scientists believe that the literary intellectuals are totally lacking in foresight, peculiarly unconcerned with their brother men, in a deep sense anti-intellectual...'

Snow gave an insightful account of this state of affairs. The problem, he thought, was that the industrial revolution was never fully embraced by the high-minded ancient universities. Even the pure (as opposed to applied) scientists failed to take it seriously for a long time. This meant that the actual machinery of industrialisation was controlled by those from the lower classes, even at a managerial level (think railways, steamships, and mines). When the engineering and applied sciences were finally admitted into the universities, the class chasm between those in the traditional faculties (classics, theology, law, etc.) and those in the new faculties simply carried over and in time became a cultural chasm. (Interestingly, this analysis doesn't see there as being an original science vs humanities divide so much as a pure vs applied one.)

As fascinating as Snow's account is, might it be missing something? Couldn't there also be a kind of *cognitive* difference between those drawn to one or the other of these two types of intellectual activity that equally and perhaps better explains the existence of the two cultures? Plenty of research suggests there are both personality and cognitive differences between STEM and arts students.[3] I think that reflection on these differences not only has a better chance of explaining the cultural difference, it might even help to dispel the 'curious distorted image' each type has formed of the other.

The most consistent finding over many decades of studies is that STEM students tend to have more convergent thinking styles, and arts students more divergent ones. Convergent thinking is associated with individuals who look for tried and tested ways of approaching tasks or problems. Convergent thinkers are willing to forego alternative solutions that may in the long run be more efficient but in the short run introduce temporary complexity or uncertainty. The divergent or 'lateral' thinker, by contrast, is associated with just the opposite tendency—to explore multiple solutions and look for unexpected connections, even at the cost of temporary inconvenience and doubt. Unsurprisingly, the divergent thinker scores higher on measures of curiosity, creativity, originality, and imagination.

Now if we accept this as even halfway right (allowing that there's bound to be at least a bit of both in each of us), then it follows that there are, roughly speaking, two psychological types distinguishable in terms of their deftness in handling natural laws, on the one hand, and in operating within what the philosopher Wilfrid Sellars called the 'space of reasons', on the other. This distinction is reminiscent of others found in the writings of the biologist Stephen Jay Gould (who spoke of science and religion as 'nonoverlapping magisteria'), the linguist Noam Chomsky (who separates human inquiry into 'problems' and 'mysteries', of which only the former, according to him, yield to the methods of natural science), and the philosopher David Chalmers (who speaks of 'easy problems', such as those attempting purely functional explanations of phenomena, and 'hard problems', such as the so-called 'mind–body' problem).

I may be mixing and stirring from many pots here, but all these distinctions strike me as trying to get at the same thing. There's the natural world on one side, and the sociocultural world on the other. So we can posit a rough division between the natural sciences, which study the natural world (think of this as the "space of law") and all other disciplines, falling under the arts, humanities, and social sciences (encompassing Sellars's space of reasons), so that each of the latter takes up a different aspect of the sociocultural world, be it artistic, literary, historical, legal, musical, ethical, political, economic, and so on. And what the studies I mentioned seem to imply is that it may be the psychological difference between the two types of individuals that makes them more or less suited to their chosen field. But what is it about the natural world that responds so well to scientific modes of inquiry and to the gaze of the convergent thinker? The answer to this question will not only bring into sharper relief how the two types of individuals and their learning styles differ—and so why they can be expected to differ culturally as well—but, more importantly, show why a deep mutual respect between the two types of individuals is warranted. At a minimum, I hope to convince the STEM sceptic that both ethics and AI ethics are *intellectually* respectable forms of inquiry.

It will surprise few readers to hear that there's no clear cut-off point where natural science ends and all other forms of inquiry begin. But at the same time, there can be no science of economics in quite the way there can be a science of physics, chemistry, and biology. Let's agree that physics furnishes science's gold standard, chemistry the silver standard, and biology the bronze standard (this isn't tendentious). Why do they all count as sciences in a strict sense, and why does physics furnish the gold standard? At the risk of oversimplification, physics posits more or less stable entities that can enter into statements of universal law. True, there are principles like Pauli Exclusion, Heisenberg Uncertainty, and so on. But in contrast to the posits (or categories) of, say, biology, the posits of physics have an enviable fixity. Consider a typical eukaryotic cell (a posit of biology). It has a mitochondrion, a nucleus, a variegated structure. It takes in nutrients and oxygen, it metabolises matter, expels toxins, grows and divides, and eventually dies. This is nowhere near as stable as an acid molecule, much less a nucleotide, much less a carbon atom, still less an electron, and so on, as you ascend the ladder of abstraction through chemistry to the simplest posits of particle physics.

Nevertheless, the relative stability of a single cell still means you can have a science of biology in something like the strict sense. Once you get to anything as complicated as an organ, let alone an organism, a society, or a whole human culture, what with its norms and customs and rituals and taboos, all bets are off. How many general statements can you make about human societies that qualify as universal laws excluding those that are either uninformative or brook a host of exceptions? (Here's an example of an informative one which, despite its perceptiveness, can't quite be considered a law of nature: 'The great majority of men and women have lived lives of fruitless toil for the benefit of a ruling elite; the political state, whatever form it has taken, has been prepared to use violence from time to time to maintain this situation; and it has been the function of quite a lot of the state's myth, culture, and thought to legitimate this state of affairs. But some form of resistance to this injustice among those who are exploited is also bound to occur.'[4]) Of course, social sciences might employ scientific techniques (null-hypothesis significance testing, experimental paradigms, surveys, etc.), and to that extent may be considered sciences. But because they don't deal in what philosophers call 'natural kinds', they cannot hope to have anything like the same success as the natural sciences in formulating universal laws that facilitate accurate prediction of phenomena and aid in their explanation.

Consider a few concepts in economics. A 'market'. A 'rational individual'. A 'perfectly competitive market' consisting of 'rational individuals', each seeking to maximise their 'utility'. If the words in scare quotes are meant to denote analogues of frictionless planes, acceleration due to gravity, and atmospherically neutral projectile pathways—the models which employ them leave a lot to be desired. This isn't to say you can't have models in economics. But the better ones, the ones that are actually useful in guiding policy, don't pretend to apply to any but fairly specific situations. If you can't have a science for every different situation, you can't have a science of the subject matter of economics (on the strict *physics-as-gold-standard* view). Even in classical physics, the path of a projectile isn't a perfect parabola owing to atmospheric conditions near the earth's surface. But you can ignore this complication in the physicist's model because the phenomenon in view is a relatively simple one. A marketplace bearing any resemblance to what's important to an economist, by contrast, is *orders of magnitude* more complex than a golf ball hurtling through space. Unlike the golf ball, one ignores market complications at one's peril. Similar reasoning would suggest you can't have a science of sociology, politics, anthropology, or law.

This, then, is a rough sketch of how (natural) science differs from other ways of investigating the world. And we can therefore suggest that there are roughly two types of investigators divided broadly by subject matter, where the subject matter parcels up between what is describable in terms of structures that are much more and much less stable. It just so happens that because the more stable the postulates, the more scope there is for universalisability—for the discovery and formulation of general laws admitting of fewer and fewer exceptions—the more traction you're apt to get using mathematics and statistics as the language of description. And if there's any kind of causal relationship between convergent thinking and scientific and/or quantitative reasoning ability—as the studies imply—then the more the style of one's thinking is convergent, the more one may be suited to the investigation of stable structures.

Conversely, the more one is a lateral thinker, the more one may be suited to the investigation of *unstable* structures—objects with indeterminate boundaries and less familiar, satisfyingly simple and elegant shapes, such as ideologies, motivations, emotions, and strange visual and musical forms (think Shostakovich and Schoenberg over Handel and Purcell).

Mind you, just where a social scientist ends up between convergent and divergent thinking camps is anyone's guess. You can be an economist in a 'hard' way (e.g. as an econometrician), in which case you'll presumably align psychologically with the STEM folk. But you can also do economics in a 'soft' way (e.g. as a socioeconomist who relates economic developments to social trends), in which case you'll presumably have more shades of the arts student about you.[5] And of course some blessed souls are polymaths, boasting both literary and numerical strings to their bow. Indeed the best scientists *have* to think laterally and creatively if they are to break free of the rigid moulds imposed by old thinking.

Now I suppose tech types could take the high prestige of STEM subjects as a cue to look down on all other forms of inquiry, in the manner of Ernest Rutherford, who allegedly said that 'there's physics, and then there's stamp collecting'. And perhaps because there always seem to be high-paying jobs in STEM, tech types may easily confuse their higher labour market status for higher overall value to society. But this would be a mistake (in fact, for more reasons than I have scope to discuss here).

The trope of the STEM genius permeates popular culture, seen in television sitcoms such as *Doogie Howser M.D.* and *The Big Bang Theory*. I suspect that many of us unreflectively feel that mathematical and natural science types are the real geniuses. Brilliant playwrights and musicians and historians may be amazingly skillful and talented individuals, but there's simply no contest when any of them are up against figures like Albert Einstein and Stephen Hawking for raw intelligence. But on reflection, this seems wrong. Although I myself am through temperament and education *parti pris* in my love for the sciences, I'm willing to bet that real insight, perceptiveness, and acuity in navigating the sociocultural sphere are *just as rare* as the skills required to be adept in advanced mathematics.

What makes us think otherwise, I imagine, is that everyone has to talk, invest in relationships, and immerse themselves in a surrounding culture. The ubiquity of social entanglements inspires the thought that because everyone *has* culture, but not everyone can do advanced mathematics, those who can do both have one up on everyone else.

But just because everyone is enculturated—most of us can read novels, watch films, enjoy a good comedy skit, and pine at hearing Elgar's *Nimrod*—it doesn't follow that everyone is or *can be* enculturated to the same degree and in the same way. It takes a certain type of person to appreciate viscerally what Shakespeare is doing in *King Lear*. It takes a certain type of person to observe, as Oscar Wilde did, that when Rubinstein plays Beethoven, he gives us not merely Beethoven, but also himself, and so gives us Beethoven absolutely. It takes a certain type of person to understand, as Umberto Eco did, that Wilde gave the world a genuine paradox when he remarked that *only the shallow know themselves*. In other words, while everyone must to some measure be immersed in culture, the advanced mathematician doesn't have the intellectual upper hand, unless the mathematician, too, can feel great works of art as deeply as the one whose life is given over to them.

The flipside of this is that just as we must all to some shallow degree engage in culture, we must all to some shallow degree operate in a technical sphere—even though most of us couldn't hope to compete with the whippersnapper at the Apple genius bar. Most of us can do arithmetic. Most of us engage in projects that require budgeting, forecasting, financial management, and navigating a healthcare system when unwell. Recent work in psychology even suggests that most of us are actually pretty good Bayesian reasoners too, possessing an intuitive grasp of probability that only *seems* to go astray under exam conditions. The matter is contested, but most of us probably aren't guilty of the so-called 'base rate fallacy' in day-to-day life. The charge is that when judging the probability of a phenomenon like an illness that's present in a tiny portion of the population, we tend to forget that tiny 'base rate' and overestimate the probability.[6] But it turns out that success in most tests of rational reasoning comes down to the precise language employed and how the problem is posed. Which of course brings us to language—a highly complex form of coding we all have to assimilate and which permits of degrees of sophistication. Again, it's no answer to say that the STEM graduate can do all this *plus* advanced mathematics; for Snow's 'literary intellectual' could just as tritely respond that the arts graduate can do everyday science *plus* ballet (or whatever form of high culture you prefer). My hunch is that we can all do STEM about as well as we can all be creative and perceptive observers of culture: we can all be technical at a level comparable to that at which we can all intuitively fathom the human condition.

None of this should be surprising. There's a basic scientific cast of mind evident even in the humanities. If you take the writing of narrative history, for example—say the epic, romantic, and emotionally charged sort associated with histories written in the grand manner of the nineteenth century—that, too, deals with large, unifying themes, patterns of cause and effect, and that which holds sway above the flux. The production of such consummate works of art displays nothing if not a scientific propensity to simplicity and parsimony. It calls to mind Nelson Goodman's remark that 'To seek truth is to seek a true system, and to seek system at all is to seek simplicity.'

And it's not just history. There's a lot in *ethics* that can be considered scientific in roughly the same way—though this isn't to say that historians and ethicists are as gifted as natural scientists at theorising over stable structures. But setting moral platitudes aside, ethical thinking is often difficult and it can be so in at least two ways, one technical, the other non-technical. Take the technical difficulty first. When someone is tempted to think that AI ethics is a soft subject, and that we should save our awe for those who build the actual technology, it's clear they've misapprehended the bulk of the work that actually goes on in ethics, including moral and political philosophy, and in the moral and political philosophy of AI in particular. A fully systematic engagement with ethics, which includes metaethics and moral psychology, requires mastering a lot of interesting work not just from anthropology and sociology, but also from behavioural ecology, cognitive archaeology, cognitive psychology, and various mathematical approaches to human behaviour, including game theory and signalling theory. Cooperation problems are, after all, coordination problems, and in trying to crack the puzzle of how cooperation evolved in our species, insights from game theory, for example, have proved invaluable.

To give a sense of the relevance of mathematical ideas in *AI* ethics, it's worth mentioning the various attempts that have been made to code fairness into algorithms. Such an effort obviously requires a mathematical criterion of fairness. But it turns out that among the many ways you can specify that metric, they're almost all incompatible, so that, if you try to optimise for one, you fall foul of another. For example, I could make an algorithm fair by ensuring that the algorithm achieves a similar proportion of false positives across different demographics, or I could make the algorithm fair by ensuring that a 70% risk of defaulting on a mortgage (say) indicates the same likelihood of defaulting regardless of demographic. The ethicists advising in this area have to be able to follow the thrust of the various arguments for and against the different fairness metrics, which in turn requires a solid grasp of the mathematics behind them. The ethical question (which criterion is fairest?) is embroiled in the technical question (how do you code fairness into an algorithm?).

But perhaps more revealingly, ethics is hard in a non-technical way. Being able to analyse a set of facts, get to the heart of the moral issue, then resolve that issue in a way that most respects everyone's intuitions about what a fair outcome would be is an intellectual skill that almost certainly requires lateral and counterfactual thinking skills.

The best example of how intellectually demanding ethics can be is legal adjudication. Being able to screen out the noisy and often very complex factual details of a case (the search for pattern amid flux), home in on the essential facts, put them in their proper sequence, isolate the relevant legal issue, then apply the law to these circumstances in a way that's most faithful to the spirit of the law, *while trying to be just at the same time*, is an intensely taxing activity. Doing so efficiently and accurately is even more so, requiring powers of attention and cognitive endurance that most people simply don't have.

For all we know, the deliberative skills required may be about as (un)evenly distributed in the population as mathematical and other STEM skills, with no necessary correlation between the two sets of skills. One of Australia's finest judges, Justice Frank Kitto, was reputedly poor at mathematics. Justice Evatt, on the other hand, himself a brilliant judge (at 36, the youngest ever appointed to Australia's highest court), was quite a capable mathematician, earning prizes in the subject while at university. In his crowning achievement while on the bench, he found in favour of a mother who suffered intense mental agony (and ultimately took her own life) after witnessing the body of her dead child being fished out of a trench of water that had been negligently left exposed by the local authority on a street in which children were known to play. The law that bound Australia at the time was against him, because in the 1930s it wasn't yet accepted that someone could recover damages for purely psychiatric harm (for what today would be called post-traumatic stress disorder). Nevertheless, in a searing and yet scholarly dissertation on the law, Justice Evatt was able to beat a path to justice, albeit in a dissenting opinion that would prove instrumental in changing the law four decades later. He managed this beset all the while by the medical orthodoxies of the time, the 'devoutly masculine sentiments' of his judicial colleagues, and the apparent applicability of an English precedent which not even the House of Lords had always seen fit to apply in the United Kingdom.[7] This is lateral thinking *par excellence*.

So rather than be stinting in our admiration for them, the STEM sceptic has reason to respect and even revere the penetration of those who come to their answers in less formal ways, who can take short cuts as if by some miracle of creative intuition. Contrary to the view among many of Silicon Valley's princelings, just because the subject isn't mathematics, doesn't mean it isn't real.

Notes

1. George Orwell, 'Wells, Hitler, and the World State', in *Selected Essays* (Oxford University Press, 2021), 115–20, 117.

2. C. P. Snow, *The Two Cultures* (Cambridge University Press, 1959).

3. For a brief review, see Adrian Furnham and John Crump, 'The Sensitive, Imaginative, Articulate Art Student and Conservative, Cool, Numerate Science Student: Individual Differences in Art and Science Students', *Learning and Individual Differences* 25 (2013): 150–5.

4. I am adapting this from lines in Terry Eagleton's *Why Marx Was Right* (Yale University Press, 2011).

5. Even at the hands of such famous twentieth-century economists as John Maynard Keynes and John K. Galbraith, most of the exposition is purely literary and discursive rather than deductive and mathematical.

6. Tomás Lejarraga and Ralph Hertwig, 'How Experimental Methods Shaped Views on Human Competence and Rationality', *Psychological Bulletin* 147/6 (2021): 535–64. Christin Schulze and Ralph Hertwig, 'A Description–Experience Gap in Statistical Intuitions: Of Smart Babies, Risk-Savvy Chimps, Intuitive Statisticians, and Stupid Grown-Ups', *Cognition* 210/104580 (2021): 1–16.

7. Gideon Haigh, *The Brilliant Boy: Doc Evatt and the Great Australian Dissent* (Simon & Schuster Australia, 2021).

Further Reading

T. Benton and I. Craib, *Philosophy of Social Science* (Bloomsbury, 2023).

19

Artificial General Intelligence

Shocks, Sentience, and Moral Status

Peter Millican

The ambition to create an *Artificial General Intelligence* or *AGI* has been around at least since Alan Turing introduced his famous test for machine thinking in his 1950 paper 'Computing Machinery and Intelligence'.[1] This test was based on the ability to simulate human-style textual conversation, potentially over a wide range of topics, so as to be indistinguishable from a real person. But subsequent attempts to pass that test were, until very recently, confined to relatively crude chatbots following the style of Joseph Weizenbaum's *ELIZA* of 1966, which emitted ready-prepared outputs in response to specific prompts (so, for example, mention of one's mother might meet with the response, 'Tell me more about your family'; and 'I need X' with 'What would it mean to you if you got X?'). Such chatbots were interesting toys, but could not plausibly be described as genuinely *intelligent*. That situation has changed radically with the arrival of *ChatGPT*, which many people see as potentially leading to a genuine AGI. And this is not only of theoretical interest, because it raises a host of moral questions, including whether such an AGI would itself be worthy of moral consideration.

The Deep Learning Revolution

Although serious research in AI had previously been fruitful in many different directions, the actual achievement of 'general' AI seemed very distant until around 2015. Then a new program, *AlphaGo*—developed by the company *DeepMind* using 'deep learning' techniques—burst onto the scene, defeating European champion Fan Hui 5-0 in a match of Go, a game which until then had been considered too subtle and complicated for computer algorithms to master in the foreseeable future.[2] The following year, *AlphaGo* sensationally went on to defeat World Go Champion Lee Sedol 4–1. And in 2017, a more generalised program, *AlphaZero*, which was capable of teaching itself new games—without any input from human experts or game databases—was able to defeat the world champion chess program *Stockfish* after only a few hours of self-training (i.e. playing games against itself). This was 20 years after *Deep Blue* had defeated Garry Kasparov, then human world champion and arguably the strongest chess player of all time. But programs such as *Deep Blue* and *Stockfish* had been specifically programmed to play chess, with input from expert players and using algorithms that were finely tuned to reflect established theory. *AlphaZero* was an altogether more impressive system, apparently able to learn new skills 'from scratch', and thus representing massive and potentially frightening progress towards Artificial General Intelligence.

Part of the promise and threat of deep learning lies precisely in its ability to represent all kinds of information in ways that it works out for itself in response to training data and feedback. This information is stored implicitly within the links of a neural network that consists of layers of artificial neurons. The input layer of the network is set up to reflect the specific problem case, so for example the activation level of the input neurons might represent the state of the squares in a chessboard position (which is to be assessed), or the colour of the pixels in a digital image (which is to be classified). The output layer is set up to signal the corresponding solution, respectively the chess assessment (e.g. how good the current position is for White) or the image label (e.g. 'cat' or 'dog' if the task is to classify images of pets). But in a deep network there may be many intermediate layers, with patterns of activation passing down through the layers from the input layer to the output layer. Beyond the input layer, the activation of each neuron depends on the inputs that it receives from the neurons to which it is connected in the previous layer. These inputs depend both on the level of activation of those previous neurons, but also—crucially—on the weight given to the relevant connection. It is these weights that are adjusted during the learning process, which typically involves going iteratively through the training data, assessing the resulting outputs and gradually refining the relevant weights until a sufficient match between inputs and outputs has been achieved. Neither the weights, nor the roles of the neurons in the intermediate layers, are predetermined when the learning process starts. By the end, the immensely complex pattern of weights implicitly represents what the network has learned, but in a way that unaided humans will find impossible to interpret, and whose behaviour they will be able to predict only inductively (i.e. by experience).

All this can look almost magical, enabling a deep network to solve problems that we have no idea how to address, and using calculations that are beyond our comprehension even when they have been discovered. But we should not be misled to suppose that the machines themselves 'understand' what they are doing in any reflective way—and not only because they are completely non-sentient (and hence have no awareness or conscious understanding of anything). For their way of working is far more closely analogous to our own unconscious pattern-recognition than it is to how we think when explicitly calculating or reasoning about something. And we are unlikely to consider that we really understand *how* we ourselves operate when recognising an animal as a dog, or a symbol as the letter 'f'—we too learn these things by examples, building up neural structures of which we are completely unaware, and whose results we find hard to articulate. Likewise in expert chess-playing— often considered a paradigm of explicit calculation—many moves are decided by a sort of trained instinct, involving implicit recognition of a position's characteristics and appropriate strategies, to the extent that a grandmaster will often choose a move with negligible calculation, and if asked, may initially have little more to say than 'in this sort of position, that is the right move to play'. (He could probably expand on the strategic significance of the move given longer to think about it, but that deeper reflection would not have featured in his move choice within the game.) There is an obvious analogy here to a neural network's method of position evaluation by pattern recognition. Of course, the grandmaster also needs to be able to tell when explicit calculation is needed, for example, when tactical combinations are imminent, but *AlphaZero* too can perform explicit calculation, exploring options down the 'game tree' of branching possibilities. This tree-searching ability was built into it, along with the rules of the game—the initial position, how pieces move, what counts as a win, draw, or loss etc. So in this sense it did not really teach itself chess *entirely* 'from scratch', despite its lack of access to human chess experts. Within that closed, rule-governed context, it was able to generate its own training data and feedback by playing lots of different games against itself, learning *by experience* which patterns were most conducive to success.

The Practical Shock of *ChatGPT*

Since 2022, we have witnessed a new 'AI shock' which seems even more significant and widespread than any that has come before, with the release of *OpenAI*'s chatbot system *ChatGPT*. This again has been developed using techniques of deep learning, but this time with colossal human textual input in the form of around 300 billion words taken from various sources on the internet. And now we suddenly—and unexpectedly—are faced with a technology that seemingly has the potential to pass the 'Turing Test' in its full generality: to converse plausibly, flexibly, coherently, and informatively about a vast range of topics, and without relying on pre-prepared outputs.

The arrival of *ChatGPT* raises major foundational questions about AGI, to be discussed below. But first, we should note some serious practical worries, with significant moral implications that would deserve extended discussion in themselves. The answers that *ChatGPT* generates look very plausible as humanly composed text, but they are unreliable, not fact-checked, and frequently involve 'hallucination' (i.e. invented claims). They are also likely to reflect biases in the source texts, and can be deliberately manipulated by asking *ChatGPT* to produce arguments for falsehoods that serve one's purposes. The system's ability to turn out believable and varied disinformation and propaganda, almost instantly and at minimal cost, provides obvious opportunities for those wishing to manipulate the online environment, for example by flooding discussion or review sites with submissions that all support a particular point of view, and yet give the appearance of having been written independently.

Related concerns arise from *ChatGPT*'s ability to generate text in specific styles as requested by the user, and thus to impersonate prominent individuals or organisations, bringing risks of fraud and deceit. That ability can be exploited more positively in the drafting of standardised forms of text, for example legal documents, though this raises serious concern about potential job losses, as humans are replaced in various routine capacities. Such replacement could in turn lead to an over-dependence on automated systems and a lack of critical perusal as the relevant processes—optimised for cost rather than quality—cease to involve reflective and suitably trained humans. This could be economically devastating for entire categories of human workers, the same workers whose online contributions over the years have been plundered by *ChatGPT* to produce its indirectly plagiarised outputs. Such developments also risk undermining accountability, as important decisions are effectively outsourced to an automated system whose workings are inscrutable and subject to numerous other concerns (as just explained). Modern companies, often intensely competing with their rivals, are systematically incentivised to reduce costs and increase revenue, and they are typically partitioned into units that also focus individually on their own budget and bottom line, with little role within these processes for broader societal concerns. So there is a significant danger that even companies that ostensibly have the best of intentions, and with employees who share those ideals, will end up following paths that optimise their short-term competitive and financial benefits, while sleepwalking progressively into a situation which nobody would have wished for.

Is *ChatGPT* an AGI?

Putting these practical concerns to one side, how should we assess *ChatGPT* and its (fast multiplying) cousins from a theoretical point of view? The system was developed by applying statistical analysis on those 300 billion words of textual data to create a *Large Language Model* (LLM) which—to simplify somewhat—records the probability that any given sequence of words (within its broader context) will be continued in different ways. And accordingly, *ChatGPT*'s primary method of working is to predict which individual words are most likely to follow in any particular textual context, and then to choose one of these words (but not always the most likely word, since an element of randomness enables it to respond differently if the same context is repeated). That chosen word is then added to the existing text, and the next word chosen in the same way, and so on. Alongside the purely automated learning which generated the (reportedly) trillion or so parameters—that is, stored probabilities—in the LLM (reflected in the deep network weights, as explained earlier), human users improved the system's responses through supervised fine tuning, in which it was given a wide range of typical prompts (i.e. potential user inputs), together with suitable human-crafted responses. Then a stage of reinforcement learning from human feedback was applied, whereby the system generated a range of responses for each given prompt, and humans assessed the relative suitability of those responses. This feedback was then statistically analysed to generate a *reward model* (i.e. a rough method of calculating the quality of response outputs), which could in turn be used to give positive or negative feedback to the network, thus refining its selection of appropriate responses.

This is, of course, a hugely simplified description of how *ChatGPT* was created and how it works. But two points here are particularly worthy of emphasis. First, its method of working is based entirely on imitating—with variation—the sorts of responses that occurred in the massive textual resources that were used to train it. Secondly, the information which it stores implicitly within its trillion or so network weights has been tuned to reflect *the characteristics of the textual data*, rather than *the characteristics of whatever domain the text might concern*. Hence, for example, if you play chess against *ChatGPT* using a common opening (e.g. a main line Sicilian Defence), it will initially respond with some of the same sensible moves that dominate its training data, and thus give the impression that it understands what is happening on the chessboard. But it stores no internal model of the board position, has no mechanisms of analysis or 'lookahead' (unlike *AlphaZero*), and is therefore subject to absurd errors, such as overlooking obvious captures, or allowing your bishop to capture one of its own pieces even when the diagonal between them is blocked. Play a less common opening, and the problems will appear much more quickly (e.g. losing track of a pawn on the fourth move of a King's Gambit). Ask it to solve a simple chess puzzle—for example, to devise a position in which White's king is on h1, Black's king on h8, and White, with only one further piece on the board, is able to checkmate Black by playing that bishop to b2—and it will have no clue. Perhaps before long these particular foibles will be dealt with, by linking up *ChatGPT* to a dedicated chess engine that really does have a model of what is happening on the board together with relevant analytical algorithms. My key point here, however, is that its apparent *general intelligence*—its ability to converse 'intelligently' about a vast range of topics where it has not been augmented with specific assistance—is based on the illusion that it has some internal understanding of what is being discussed. But unless the language model in some domain is able to generate indirectly a reliable model of the reality (which looks most plausible in domains that are primarily conceptual or expressive), it will have nothing like such internal understanding, and can quite appropriately be described as a 'stochastic parrot'.[3]

The hype surrounding *ChatGPT*, and failure to appreciate the points just made, could lead to real dangers if humans come to depend on its outputs by assuming that they have been generated through some kind of genuine understanding. One obvious area of concern is software development, where the apparent ability of *ChatGPT* to perform impressively in programming exercises might tempt companies to rely on it for code production, and thus hugely economise on their software engineering teams. But as in routine chess openings, the reason why *ChatGPT* does so well in those exercises is not that it rigorously analyses some internal model of what is going on in the code that it generates, but rather, that its own implicit textual model is based on analysis of the many millions of programming examples that it has been given, and it accordingly generates the code by mimicking, with what appear to be appropriate textual variations, the dominant patterns. So if you have a programming task that is very standard, or which deviates from a standard task in relatively common ways (e.g. where teachers have devised non-standard programming tasks for their students, such as sorting an array of numbers but with odds preceding evens), then *ChatGPT* might well come out with a correct solution. But if the task involves any nuances (or combinations of factors) that are less common or entirely novel, then there is a serious risk that you will end up with code which has the double disadvantage of looking very plausible—and quite possibly works well in most cases—whilst actually being incorrect, thus making the errors especially hard to identify.

None of this is to deny that in expert hands, *ChatGPT* can be a useful programming assistant, providing quick and targeted search through the mountains of online programming resources to generate 'first draft' answers that are often on target. But the use of *ChatGPT* in programming by those who are unaware of its foibles, or insufficiently expert to check carefully what it produces, carries the risk of generating false confidence, both in the code that it generates and in the programmers who rely on it (whose lack of understanding could be masked by their success in turning out plausible code). Empirical studies consistently show that far more programmer time is taken fixing 'bugs' than writing original code, so focusing on saving time on the latter, at the cost of code quality, is a dangerous false economy. Companies whose managers are taken in by the hype of supposed AGI could easily find themselves relying on large amounts of parroted code that reflects limited understanding of the problem at hand, and which, when it goes wrong, is either a nightmare to debug or needs to be replaced entirely. This is a serious potential danger, because the combination of poor software quality and excessive confidence in its reliability has notoriously led to many public scandals, involving huge personal and commercial loss (for example, the Horizon IT scandal which led to numerous unjustified prosecutions of British sub-postmasters, resulting in wrongful imprisonment, bankruptcy, and even suicide).

Could an AI System be Sentient or Worthy of Moral Consideration?

Despite all these reservations and limitations, it is understandable that the development of Large Language Models has triggered huge interest and concern around the possibility of AGI. *ChatGPT*'s ability to converse so plausibly over such a wide range has come as a tremendous surprise to most people, including AI researchers, and the long-standing influence of the Turing Test has led many to view such conversational competence and versatility as the hallmark of AGI.

Let us, then, imagine that in the not-too-distant future, some comprehensive AI program is designed using broadly similar technologies, enhanced where necessary (as has already started happening) with a wide range of specialist modules that enable it to handle tasks requiring particular types of detailed calculation that go beyond the powers that it has managed to learn for itself. In these cases, the conversational interface of this hypothetical program translates natural language queries into some internal form which explicitly identifies the required task. Then if the task involves chess, a dedicated chess engine will be used to find the answer, which is then translated back into natural language. Likewise, if the task involves mathematics, or programming, or whatever, the necessary processing will be done either by a specialist module, or perhaps by some more versatile system which is able to construct and manipulate representations of non-standard scenarios (e.g. to deal with arbitrary logic puzzles). In 2024, this possibility does not seem nearly as remote as it did a decade ago.

Suppose, then, that we are presented with this impressively versatile system, which seems to be capable of plausibly intelligent responses over a vast range of topics, and which—unlike *ChatGPT* and its ilk—also appears to be highly reliable over the entire range, generating answers that are overwhelmingly correct.[4] We would thus have a system which plausibly passes the Turing Test and, in at least one reasonable sense of the phrase, is an *Artificial General Intelligence*. Does this mean, however, that we would also have an AGI in another common sense of that phrase, namely, a system capable of human-level cognition in general, including that mysterious, elusive, and morally fundamental quality of sentience or consciousness? The two senses are very often confused or merged, but I shall end by arguing that they should be sharply distinguished.

Mostly, when we encounter two systems whose external behaviour is similar, we naturally and reasonably suppose that they are likely to be similar internally also. So when we see animals acting in ways that are analogous to our own primitive behaviour—for example seeking food or shelter, defending themselves, nurturing infants, or interacting with a potential mate—we suppose that they are intentionally motivated in much the same way as ourselves, either by instinctive desires or by the prospect of pleasures and pains. And if their behaviour is highly sophisticated and flexible in response to variable circumstances and interactions with other animals, we would also suppose that this motivation is to some extent controlled by conscious reasoning. We would presumably make similar judgements if we were to encounter some apparently intelligent alien life form, and it might then seem inevitable that, as Turing advocates, 'fair play for the machines' mandates our applying the same criteria to artificial intelligence systems.[5]

But to illustrate where this seductive line of thinking goes wrong, suppose that we have standard criteria for identifying a wood fire in terms of its external signs: patterns of heat and light, crackling sounds, and smoke. And now suppose that we are confronted with an electronic device which is intricately designed to mimic all of these features, using heating elements, lights, speakers, and artificial smoke. Knowing how it works, do we have good reason to apply our standard criteria and conclude that wood burning is taking place? Of course we don't, because we know that those standard criteria are being fulfilled in a *different* way using clever electronics rather than real combustion, and we are certain that no wood at all is involved in the process.

A parallel point can be made about our supposed AGI, for we know at least in outline how it is generating its 'intelligent' responses, using sophisticated electronics implementing deep neural network structures, fed with a massive textual corpus and tuned by clever machine-learning algorithms that have been designed precisely for the purpose. We might well be somewhat mystified as to how exactly the finished system works in terms of all those inscrutable representations stored implicitly in the network weights, but we know in general terms how the network is operating—including a *precise* understanding of the functions of its individual digital components—and we know that all this involves no element whatever of consciousness or sentience on the part of the machine. So to suppose that somehow the system magically becomes conscious once some threshold of complexity has been crossed (in the way beloved of sci-fi films such as *Terminator*) is complete fantasy.

But if this argument works, a critic might ask, why would it not apply equally to other humans, albeit they are made of *biological* (rather than *electronic*) hardware? And would that not lead to solipsism (as Turing alleged in his 1950 paper, §6.4)? One obvious reply is that we emphatically *do not* have a complete functional understanding of our own biological hardware: brain science is at a very early stage, and its biochemistry is extraordinarily complex (not to mention that our understanding of basic physics remains fundamentally incomplete). So we have no good reason for assuming that the behaviour of our neurons can be abstractly modelled in a similar way to digital components, let alone that they could adequately be replaced with electronic substitutes (as is commonly assumed in fanciful thought-experiments). Another obvious reply is that we are immediately aware of our own consciousness as something which is more than merely an abstraction, and is causally efficacious (e.g. we seek things that give us conscious pleasure, and shun things that give us pain). This causal efficacy, indeed, is key to its evident evolutionary tuning, whereby conscious pleasures and pains are strongly associated, respectively, with things that are evolutionarily beneficial and harmful (a coincidence which would otherwise be wildly improbable). This gives an excellent basis for both inferring consciousness in other beings that have evolved in a similar manner, and for asserting that our own mental processes cannot be reducible without remainder to the digital functions of logic gates etc., which may combine to generate sophisticated *abstract* processing (such as calculation of chess moves), but cannot generate a novel *causal* mechanism that goes beyond those functions.[6] We do not know how our consciousness is caused by our biological hardware (or, some would say, by an immaterial soul), but however it may be caused, we can be confident that it is not purely abstract.

This case against the possibility of conscious digital machines is not particularly sophisticated, and may appear to many as plain common sense. But I suspect that it is often overlooked precisely because of the unfortunate conceptual blurring implicit in our notion of 'intelligence'. On the one hand, we tend to judge intelligence in terms of efficient and effective information processing in the solution of relatively complex problems. On the other hand, we tend to presume that intelligent entities will be purposively motivated by conscious goals, devoting their energies to solving the sorts of problems that stand in the way of their achieving those goals. Since Alan Turing's momentous invention of his 'computing machines' in 1936, however, we know that these two aspects can come apart: it is entirely possible to have efficient and effective information processing without any trace of sentience, and it is likewise possible to have systems—for example chess computers—that behave purposively to achieve goals without those goals being the least bit conscious. Thus it is perfectly reasonable to allow that such systems are exhibiting genuine intelligence, in the sense of *sophisticated information processing to good effect*, without in any way having to accept that this implies genuine *sentience*. Indeed, we have excellent reason for denying that the latter is even possible for our artificial electronic machines, because we know (in general terms) how they operate, and this removes any justification for such fanciful speculation. So even while our own conscious awareness and its biological basis continue to be fundamentally mysterious to us, this gives no reason for extending the scope of that mystery so that we come to wonder about the potential sentience of our electronic computer systems. Their non-sentience is something of which we can be certain, precisely because we know how they operate right down to the digital functions of their most basic computational components, and we also know how those functions give rise to the intelligent behaviour. This is an argument based on positive knowledge, not ignorance.

As for whether such artificial systems are worthy of moral consideration (i.e. as moral *patients* rather than *agents*), my answer would be an emphatic 'No!' Philosophers such as Immanuel Kant may have argued that moral value is a matter of *rationality*, but that always seemed implausible to those of us who took animals (and Darwin) seriously—it is far more plausible to see moral consideration as applying to creatures that are capable of *feeling*, and here David Hume rather than Kant points the way. Indeed, this position is significantly strengthened by the creation of artificial intelligences which can perform rational information processing but lack any inner life. Utterly incapable of real desires, pleasures or pains, fulfilment or suffering, it follows that they also lack any intrinsic interests. Hence such artificial intelligences cannot possibly be worthy of moral consideration.

Acknowledgement

I am very grateful to Vince Conitzer, David Edmonds, Mark van der Wilk, and Michael Wooldridge for helpful comments on drafts of this paper.

Notes

1. In the journal *Mind* volume 59, pp. 433–60. Turing himself did not use the term *Artificial General Intelligence*, which started to become commonly used around the turn of the century.
2. For a vivid and accessible account of the history of machine learning and the beginning of the deep learning revolution, see pp. 167–200 of Michael Wooldridge, *The Road to Conscious Machines: The Story of AI* (Penguin, 2020).
3. The term 'stochastic parrot' was coined by Emily M. Bender in 2021. Recent work by my Oxford colleagues Philipp Koralus and Vincent Wang indicates that the most recent generations of *ChatGPT* have become better at mimicking

good human thinking, but also parroting *fallacious* human thinking (arXiv:2303.17276v1). Prompt engineering and generation of synthetic data-sets might mitigate this weakness, but again, such methods focus on enabling the system to find answers through modelling of *human language*, rather than of *the relevant domain*.

4. Note that this might be very hard to establish. Even if the system responds brilliantly well in the large range of situations that have been anticipated (and perhaps "patched" with specialist modules), it could still remain the case that when something is asked which "falls between the cracks", the system will be utterly—and perhaps dangerously—useless. This contrasts with typical human performance, where we are able to infer with confidence that someone who is excellent in general will not be utterly stupid beyond their expertise. So we need to be very cautious about extrapolating such expectations to AI systems.

5. For appreciation of how far the Turing Test setup was motivated by a desire for such fairness—rather than primarily by seeing human indistinguishability as a *criterion*—see the article cited in the Further Reading, pp. 37–8, 45–6. See also pp. 47–9 for further discussion of the issues raised here.

6. This point applies even if we have an AI system which is itself 'evolving' algorithmically in response to feedback. All such 'evolution' involves specific changes in the operation of a system whose basic processes we understand well, and which we know to involve no element of consciousness. By contrast, it seems that biological evolution was able somehow to exploit properties of matter which we do not currently understand, and which generate distinct causal mechanisms associated with consciousness.

Further Reading

Peter Millican, 'Alan Turing and Human-Like Intelligence', in Stephen Muggleton and Nicholas Chater (eds), *Human-Like Machine Intelligence* (Oxford University Press, 2021), 24–51, especially pp. 36–49.

20

Human in the Loop!

Ruth Chang

In the not-too-distant future, technology might help us decide which school to attend, what career path to pursue, and whether to marry and have children. It might also help our governments decide how much to tax its citizens, which regulations to impose on businesses, how to manage energy consumption and distribute healthcare resources, whether to go to war, and even how to manage deep political disagreements. We should, I think, welcome this decision-making assistance.

One way machines could help us is as a mere aid or tool, much like a calculator helps us decide what tip to leave at a restaurant. When deciding between a career in law and graphic design, for instance, a career App of the future might tell us the percentage of people, similar to us in psychological and cognitive profile, who have succeeded as lawyers compared to those who have succeeded in graphic design. We might then use this information as part of our deliberation about the pros and cons of pursuing each career. At the end of the day, however, we would make the decision in the old-fashioned, human way, with technological output as nothing more than an informational aid.

Today technology is typically deployed in just this way. Machines tell us that certain features highly correlate with being a successful employee; that having a certain profile is predictive of having exorbitant lifetime medical costs; that living in a certain postal code makes it more likely than not that a person will default on a loan; that granting bail to a defendant with such-and-such features involves a certain chance that they will commit another crime while free, and so on. What an employer, healthcare policy-maker, loan officer, or judge does with this machine-generated information, however, is up to them. They can treat it as settling the matter at hand, ignore it all together, or something in between.

Although this use of technology—as a mere aid to human decision-making—appears relatively tame, it has some significant downsides. The most notorious is that technologies—for example, facial recognition, machine learning, natural language models— give us biased outputs, reproducing our own prejudices and bigotries since they train on data provided by flawed human attitudes and behaviour. If we treat such outputs more like prophecies than the off-colour musings of a retrograde relative, we are likely to end up exacerbating unfairness at a systemic level. The problem of bias and unfairness, perhaps in conjunction with a general (and surprisingly resilient) fear that current technology is a stepping stone to an artificial *general* intelligence that could in principle obliterate us, has led some technologists to call for a red line to be drawn in the silicon: we should create technology that, *at best*, operates as a mere aid to *but never as a substitute for* human decision-making. Allowing machines to determine all by themselves what we humans should do—what career we should pursue, whom we should hire, whether we should grant a defendant bail, etc.—crosses a line and should be strictly prohibited.

This attempt to ring-fence technology strikes me as unfortunate for two reasons. First, whatever its wisdom, it is doubtful that any government or organization *could* ensure—certainly not globally— that technology by and large operate as mere aids to, rather than substitutes for, human decision-making. Second, the horse has already bolted: technology that operates as a substitute for human decision-making is already on the scene. Social media algorithms already decide what news we read and banking algorithms have for years determined whether we will qualify for a mortgage. We should, I believe, accept what appears to be inevitable: that technology that operates as a substitute for human decision-making—that is, technology that makes choices for us—will continue to be built and, indeed, proliferate. And so we should focus our energies on making such technology as useful and safe as it can be.

One natural route to building useful and safe decision-making machines involves *aligning* human and machine values. Alignment is arguably the most important open problem in AI.[1] Today, the leading strategy for attempting to achieve alignment is to 'put the human in the loop' of machine processing. By requiring human input at critical junctures of machine processing, we can—so the hope goes—bring machine decision-making in line with human values.

This idea of putting the human in the loop has a variety of implementations. In autonomous weapons, for instance, humans are 'in the loop' in that they must *initiate* such weaponry; a machine is not allowed to decide by itself whether to bomb a village. Sometimes humans are in the loop when they have the power to abort a machine's output; if a human doesn't approve of a machine's decision to use a cluster bomb, it can *abort* that output (this is also called being 'on the loop'). There are more interesting ways humans can be in the loop, too. If a machine-learning algorithm is uncertain about its reward function, it can observe or interact with humans and thereby—in principle—learn the human's reward function.[2] In mixed machine learning rules-based programming, a human might input specific weights to be assigned to factors in decision-making.[3]

This chapter sketches a novel way of putting the human in the loop of machine processing. The key to the proposal is to design technology so that the machine recognizes *hard choices*. Hard choices are between alternatives that can be compared, and yet neither is better than the other and nor are they equally good: they are *on a par*. Typically, alternatives are on a par when one is better in some respects, the other is better in other respects, and yet neither is at least as good as the other overall. They are qualitatively different from one another, and yet in the same neighbourhood of value overall. Should you have children or devote yourself to a career? Should you be a physician or architect? Live in the country or city? We humans face hard choices all the time. If machines are to both align with our values and make choices for us, they should face hard choices too.

Surprisingly, current technological design makes no room for hard choices. Indeed, the philosopher and AI expert Bryce

Goodman has argued that the existence of hard choices in human life places a 'hard limit' on building decision-making AI.[4] While some of the most sophisticated technological design makes room for uncertainty, incompleteness/incomparability, and indeterminacy,[5] hard choices are a distinct phenomenon.[6] They require a new approach to AI design.

If you face uncertainty, incomparability, or indeterminacy, it is always intrinsically permissible to arbitrarily select between options. By 'intrinsically' I mean 'on the basis of how the options relate to one another', in contrast to 'extrinsic' bases, such as what would make you most popular or save you the most time. If your choice is shot through with uncertainty, it's as if you are choosing between two black boxes, and it's permissible to flip a coin between them. If your options cannot be compared, they are outside the scope of rational choice and, once again, it is permissible for you to randomly select between them. And if it's indeterminate which option you should choose because a relevant concept is vague, you are permitted arbitrarily to tighten up that concept one way rather than another to settle the matter.[7] Thus, a machine confronting uncertainty, incomparability, or indeterminacy in its options would always be intrinsically permitted to flip a coin. But in hard choices it is *never* intrinsically permissible to flip a coin to settle which to choose.

Critically, in hard choices, a machine should halt its processing and await *human* input, which then alters its processing going forward. I believe that putting the human in the loop at the juncture of hard choices takes us closer to solving machine–human value alignment than any existing alignment strategy. This is because recognizing hard choices—as distinct from uncertainty, incomparability/incompleteness, and indeterminacy—corrects a fundamental mistake about value made in current technological design.

At present, if you ask a machine whether you should choose A or B, it is limited to only three possible positive outputs: 'choose A!', 'choose B!', and 'flip a coin!', where flipping a coin might be the output because A and B are equally preferred, incomparable, or there is uncertainty or indeterminacy about which one should choose. These three possible positive outputs correspond to three possible ways in which A and B can be related: A can be better than B, worse than it, they can be equally good. Current technological design simply assumes that this trichotomy of possibilities exhausts the conceptual space of positive ways in which alternatives can relate. If instead we design technology to make room for hard choices, there will be not three but *four* possible positive outputs: 'choose A!', 'choose B!', 'flip a coin!', and 'hard choice!' If the choice is hard, that is because the options are *on a par.* Parity, distinct from uncertainty, incomparability/incompleteness, and indeterminacy, is a fourth basic positive way in which alternatives in choice can relate. Value is not trichotomous, as technological design currently assumes, but *tetrachotomous* in structure.

When we humans face hard choices, we can do one of two things. First, we can *commit*, that is, put our very selves (or an entity we have the authority to represent) behind some feature of an option that then adds value to that feature, typically making the option more valuable than it was before.[8] The choice then becomes easy; we should choose the better option. Our 'standing for' some feature—putting our very selves behind it—endows that feature with added value.[9] In this way, *we*—our agency as such—are sources of normativity. But humans are not always ready to commit, or even capable of committing, in a given set of circumstances. Second, we can instead *drift,* that is, intentionally choose one of the options on the basis of its valuable features, but without committing to any of its features. When we drift into an option, we choose it even though it is on a par with other options, usually because we have to choose *something* and we can't quite commit to any feature of the options. Because drifting does not involve commitment, there is no change in the values at stake going forward.

An example might help. Suppose you are a marketing manager of a firm looking to promote a member of your team. You want to choose the best candidate with respect to loyalty, productivity, and team spirit. Two candidates rise to the top: Fabio, who is a consummate team player, lap-dog loyal, and an okay producer, and Ellie, who is the most productive person on the team, unafraid to call out the boss, but an awkward loner. Whom should you promote?

Neither is better than the other with respect to all the criteria that matter, but nor are they equally good—a small but definite improvement in one would not make them better than the other. They are *on a par*, and the choice between them is hard.

You might, as an exercise of your authority as a manager, commit to Fabio's quality of being a great team player. Through your authority at the firm, you commit the firm to this feature, endowing it with more value than it had before. This in turn will likely make Fabio better than Ellie. If it does, you should promote Fabio. Since the firm now stands behind this specific quality, this then affects how candidates are evaluated going forward, giving a leg up to those candidates who are similarly great team players.

But you might instead find yourself unable to commit to any particular feature of the candidates. Perhaps you *personally* could commit to the quality of being a team player, but you might think that such a commitment would be an abuse of your authority within the firm if, for example, you know that other department managers have all committed to productivity in their promotion decisions, and your position in the firm includes a duty to mirror or cooperate with the decisions of your peers. If, wearing the hat of marketing manager, you are unable to commit to any feature of the candidates, you might simply drift into choosing one candidate over the other. You might, for instance, drift into promoting Ellie over Fabio. Because you haven't made a commitment, you don't change the values at stake in the choice. And hence you don't affect the values at stake in the next case of promotion you might consider. Next year, for example, you may face another hard choice between promoting Georgina, a highly productive but surly lone wolf, and Hariri, the person everyone wants to have on their team. You might reasonably choose to promote Hariri. Had you previously committed to some feature, such as Fabio's particular quality of being a great team player, and thereby endowed it with added value, perhaps the choice between Georgina and Hariri would not have been hard. Your previous commitment could have made the choice of Hariri easy. In this way, what you do in hard choices affects the normative landscape of your decision-making going forward.

Could a machine be built that made promotion decisions for you? If we put the human in the loop in the right way, viz., at the juncture of hard choices, I believe that such a machine could be built that aligns with human values. One simple idea is to combine machine-learning algorithms, which rank candidates with respect to each criterion relevant to the choice, with rules-based programming, which puts those rankings together by some function to yield a final ranking of candidates. Such a machine could in principle be built in four steps.

We might start by hard-coding 'hard cases' into the training data of machine-learning algorithms.[10] One crude way to do this might be to use the time lapsed between hiring and promotion as a proxy for how difficult a decision it was to promote a candidate. The longer the period, the more problematic the case for promotion relative to other candidates. No doubt we could – and would need to – devise more sophisticated ways of mining data to code cases as 'hard'.

More importantly, we would, second, need to redesign machine-learning algorithms so that they recognize hard cases. This requires a new modelling for algorithms that recognizes the tetrachotomous, rather than trichotomous, structure of value. This is by far the most important step in getting machines to recognize hard choices; we need nothing short of a new decision theory that replaces the ubiquitous, trichotomous, expected utility and its cousins. Elsewhere, I have suggested a possible philosophical basis for such a model. Take as the unit of analysis not the value of items but the difference in value between them. Such differences could be represented by intervals: if one option is better than another, their approximate difference will be given by an interval of positive numbers, such as [2, 5]; if it's worse, their approximate difference will be an interval of negative numbers, such as [-1, -8]; if the items are equally good, the interval will be a special case of [0, 0], and if they are on a par, their approximate difference will be represented by an interval surrounding zero, such as [-3, 15]. These interval values could correspond to or be generated by different legitimate ways of putting together the disparate values at stake in the choice.[11] Developing these ideas with greater mathematical sophistication, philosopher and logician Kit Fine is creating a model of decision-making under parity (and risk) that could be the basis not only for new algorithms that recognize hard cases but also new tetrachotomous models of decision-making in general.[12]

With numerical representations of approximate differences in place, machines could, third, calculate the approximate difference between any two options with respect to each of the criteria relevant to choice. This might show, for example, that Ellie is better than Fabio with respect to productivity, that Fabio is better than Ellie with respect to team spirit, and that the two are on a par with respect to loyalty.

Armed with these rankings relativized to each criterion, the machine would then, fourth, be given a range of weights to be assigned to each criterion. At present trichotomous weights are usually assigned by an engineer trying to achieve plausible outputs.[13] On Fine's model, humans would ascertain the 'approximate ratio' of weights of the criteria on independent grounds; humans would tell the machine that, for instance, loyalty counts for *at least* one-tenth and *at most* one-half as much as productivity in determining the best candidate to promote.[14] And so the weight of loyalty would be represented by the interval $[\frac{1}{10}, \frac{1}{2}]$. With the weights of each criterion in hand, the machine could then begin processing candidates for promotion according to the three criteria of loyalty, productivity, and team spirit. Critically, whenever the machine encounters items on a par, that is, items whose approximate difference is an interval surrounding zero, it would halt processing and await human input.

You could in principle allow a machine to make your promotion decisions for you. Assuming all goes well, the machine would rank all candidates below Ellie and Fabio while flagging Ellie and Fabio as a hard choice. At this point the machine would halt its processing and send you a signal that it is awaiting your input.

You might respond to the machine's request for input by reviewing the dossiers of Ellie and Fabio and then committing to Fabio's team spirit. In this case, you would then send a message to the machine instructing it to adjust the ratio weighting of the criterion of team spirit the minimal amount required so that Fabio now becomes better than Ellie. The machine would make the smallest change required in the interval representation of the ratio weight of team spirit so that the approximate difference between Fabio and Ellie is no longer an interval surrounding zero but a positive interval favouring Fabio. Critically, this adjustment affects the machine's processing of candidates going forward. Next year, when you are deciding between Georgina and Hariri for promotion, the machine may find no hard choices, yielding the output that Hariri is best. If you had *not* committed to team spirit in the case of Ellie and Fabio, and had the machine *not* adjusted the ratio weight of that criterion accordingly, Georgina and Hariri would have been flagged as a hard choice. In this way, what one does in the face of a hard choice determines which choices are hard going forward.

Could the machine process a hard choice without you? Perhaps it could simply predict what you *would* have committed to had you considered the hard case, and then happily continue its processing in your absence. But if machines that make decisions for us are to have any hope of aligning with our values, the human must be in the loop at the juncture of hard choices. Note that you can't endow something with value without *actually* committing to it. Mere prediction by a machine as to what a human *would* commit to had they considered the hard choice will not do; without the human *actually* considering the hard choice and making a commitment, the machine will fall out of alignment with human values. If, for example, the machine were to process the hard case of Ellie and Fabio without you, by predicting—correctly—that *were* you to consider the case, you *would* commit to Fabio's team spirit and then adjusting the weight of team spirit accordingly, it will no longer be aligned with your values. This is because you haven't in fact considered the case and haven't in fact committed to Fabio's team spirit, and so by your lights, team spirit does not have added value whereas by the machine's lights it does. Next year, for instance, while you would consider the question of whether to promote Georgina or Hariri a hard choice, the machine treats it as easy. If machines are built to recognize hard cases and are to align with human values, humans must be in the loop.

You might instead respond to the machine's request for input in the hard case of Ellie and Fabio not by committing but by drifting, intentionally choosing one candidate on the basis of their good-making features without standing behind any of those features. Suppose you drift into promoting Ellie. In this case, you would send a message to the machine to prefer Ellie over Fabio without making any adjustments to the ratio weightings of the criteria. This output of 'Ellie beats Fabio' is treated as a special ordering that does not have implications for future machine outputs. Once again, the machine would become unaligned unless you, as authoritative decision-maker, in fact drifted in this decision.

In this way, what we actually do in hard choices—*commit* or *drift*—changes the normative landscape of the values of options going forward. Thus, if machines are (i) to make decisions for us, (ii) in accordance with our values, (iii) in the face of hard choices, they must be designed so that the human is in the loop at the juncture of hard choices.

This will, I believe, take us a long way toward solving the alignment problem.

Acknowledgement

Thanks are due to The New Institute and the Wissenschaftskolleg for financial support and to Dave Edmonds for very helpful comments and for inviting me to contribute to this project of bringing philosophical thoughts about technology to a wider public.

Notes

1. See Melanie Mitchell, 'What Does it Mean to Align AI with Human Values?', *Quantamagazine* (2022), <https://www.quantamagazine.org/what-does-it-mean-to-align-ai-with-human-values-20221213/>.
2. See Stuart Russell, 'Learning Agents for Uncertain Environments' (extended abstract), *Proceedings of the Eleventh Annual Conference on Computational Learning Theory, COLT '98, ACM*, New York (1998), 101 (1255, 103); Andrew Ng and Stuart Russell, 'Algorithms for Inverse Reinforcement Learning', *Proceedings*

of the Seventeenth International Conference on Machine Learning (2000): 663, 670; Dylan Hadfield-Menell, Anca Dragan, Pieter Abbeel, and Stuart Russell, 'Cooperative Inverse Reinforcement Learning', 30th Conference on Neural Information Processing Systems (NIPS, 2016), Barcelona, Spain; Stuart Russell, 'Provably Beneficial Artificial Intelligence', Future of Life Institute (2017), <https://people.eecs.berkeley.edu/~russell/papers/russell-bbvabook17-pbai. pdf>; and Stuart Russell, *Human Compatible: AI and the Problem of Control* (Viking, 2019).

3. I suggest a modification of such mixed programming later in the chapter.

4. Bryce Goodman, 'Hard Choices and Hard Limits for Artificial Intelligence', Proceedings of 2021 AAAI/ACM Conference on AI, Ethics, and Society (AIES '21), 19–21 May 2021, Virtual Event. ACM, New York, <https://doi. org/10.1145/3461702.3462539>.

5. See Russell, *Human Compatible*; Roel Dobbe, Thomas Krendl Gilbert, and Yonatan Mintz, 'Hard Choices in Artificial Intelligence', *Artificial Intelligence* 300 (2021): 103555, <https://doi.org/10.1016/j.artint.2021.103555>; Duncan McElfresh, Lok Chan, Kezie Doyle, Walter Sinnott-Armstrong, Vincent Conitzer, Jana Schaich Borg, and John P. Dickerson, 'Indecision Modeling', 35th AAAI Conference on Artificial Intelligence (2021).

6. For more discussion on why hard choices are not properly understood as cases of uncertainty, incomparability, or indeterminacy, see Ruth Chang, 'The Possibility of Parity', *Ethics* 112 (2002): 659–88; Ruth Chang, 'Hard Choices', *The American Philosophical Association Journal of Philosophy* 92 (2017): 586–620, <https://doi.org/10.1017/apa.2017.7>; and Ruth Chang, 'Are Hard Cases Vague Cases?' in Henrik Anderson and Anders Herlitz (eds), *Incommensurability: Ethics, Risk, and Decision-making* (Routledge, 2021).

7. Note that Dobbe et al., understand hard choices as cases of indeterminacy and suggest that democratic deliberation is required in such choices. But they seem to overlook the fact that in cases of indeterminacy, arbitrary stipulation is always intrinsically rationally permissible, and it is unclear how democratic deliberation could have the authority to settle a hard choice if, as an intrinsic matter, one could just as well flip a coin between the alternatives. I suggest that what Dobbe et al. have in mind by 'hard choices' are cases of parity. In cases of parity, human input—perhaps group deliberation—is required to settle the choice.

8. I say 'typically' because exactly how things shake out, evaluatively speaking, depends on a substantive theory of value. The correct such theory will recognize the possibility of organic unities; sometimes adding value or normativity to something doesn't make it better or supported by most reason. For more discussion of how commitments can be a source of normativity, see Ruth Chang, 'Voluntarist Reasons and the Sources of Normativity', in D. Sobel and S. Wall (eds), *Reasons for Action* (Cambridge University Press, 2009), 43–271; Ruth Chang, 'Grounding Practical Normativity: Going Hybrid', *Philosophical*

Studies 164/1 (2013): 163–87, <http://link.springer.com/article/10.1007% 2Fs11098-013-0092-z>; Ruth Chang, 'Commitments, Reasons, and the Will', in R. Shafer-Landau (ed.), *Oxford Studies in Metaethics*, vol. 8 (Oxford University Press, 2013), 74–113. Note that in the limiting case, committing to a feature of an option is committing to the option itself.

9. Ruth Chang, 'Grounding Practical Normativity'; Ruth Chang, 'What Is It To Be a Rational Agent?', in Ruth Chang and Kurt Sylvan (eds), *The Routledge Companion to Practical Reason* (Routledge, 2021), 95–110.

10. Strictly speaking, this first step could be optional if the other steps are executed appropriately. Hard cases could be identified even if the training data assumed trichotomy.

11. Ruth Chang, *Making Comparisons Count* (ed. Robert Nozick) (Routledge, 2002).

12. Kit Fine, *Modelling Parity and Imprecision* (unpublished manuscript).

13. For some criticism of this approach to assigning weights, see Ruth Chang, 'Does AI Design Rest on a Mistake?' (unpublished manuscript).

14. These approximate ratios of the weights of the criteria could in principle be elicited through a question and answer session with the computer.

Further Reading

B. Christian, *The Alignment Problem: How Can Artificial Intelligence Learn Human Values?* (Atlantic Books, 2020).

Index

For the benefit of digital users, indexed terms that span two pages (e.g., 52–53) may, on occasion, appear on only one of those pages.